Mike Dash is a Cambridge-educated historian and the author of several books acclaimed for their combination of quality writing and detailed original research. His titles include the international bestseller *Batavia's Graveyard*, and more recently *Thug* and *The First Family*. Dash is currently at work on another history and contributes regularly to newspapers and magazines.

By Mike Dash

Borderlands
Tulipomania
Batavia's Graveyard
Thug
Satan's Circus
The First Family

Tulipomania

The Story of the
World's Most Coveted Flower
and the Extraordinary
Passions It Aroused

MIKE DASH

WEIDENFELD & NICOLSON

A W&N PAPERBACK

First published in Great Britain in 1999
by Victor Gollancz
This paperback edition published in 2003
by Weidenfeld & Nicolson,
an imprint of Orion Books Ltd,
Orion House, 5 Upper St Martin's Lane,
London WC2H 9EA

An Hachette UK company

3 5 7 9 10 8 6 4

Reissued 2010

Copyright © Mike Dash 1999

ISBN 978-0-7538-2799-4

Printed and bound in Great Britain by
Clays Ltd, St Ives plc

The Orion Publishing Group's policy is to use papers that are natural, renewable
and recyclable products and made from wood grown in sustainable forests.
The logging and manufacturing processes are expected to conform
to the environmental regulations of the country of origin.

www.orionbooks.co.uk

For Ffion

*They were possessed with such a Rage or, to give
it its proper Name, such an Itching for their Flowers,
as to give often three thousand Crowns for a Tulip
that pleased their Fancies; a Disease that
ruined several rich Families.*

Monsieur de Blainville, *Travels through Holland*
(London 1743) I, 28

Contents

The United Provinces of the Netherlands in the Golden Age

NORTH SEA

TEXEL

Wieringen

WEST FRIESLAND

Medemblik

Enkhuisen

Alkmaar · Hoorn

ZUIDER ZEE

HAARLEM-AMSTERDAM CANAL

Kennemerland

Haarlem

Harlemmermeer

Amsterdam

Ij

HOLLAND

Leiden

UTRECHT

Utrecht

The Hague

Delft Gouda

Rotterdam

VIANEN

Brill

Dordrecht

ZEELAND

GELDERLAND

LANDS OF THE STATES GENERAL

Secured from Spain 1600-48

Antwerp

Brussels.

Maastricht

Rhine

FRIESLAND

GRONINGEN

DRENTHE

OVERIJSSEL

MILES

0   10   20   30   40   50

To OVEREEN
and BLOEMDAAL

GARDENS

BLEACHERIES

CANAL

AMSTERDAMSEPOORT

ZIJL STRAAT

GHEDESTRAAT

JANS STRAAT

MARKET
PLACE

GROTEKIRK

KONINGSTRAAT

HAGESTRAAT

BOTER
MARKT

GROTE HOUTSTRAAT

KLEIN HEILIG

KLEIN HOUT

SOPHIA STRAAT

GROTE
HOUTPOORT

N

GARDENS

WAGENWEG

DREEF HOUTPLEIN

KLEINE HOUTWEG

GARDENS

To HEEMSTEEDE

HAARLEM
At the time of
the
TULIP MANIA

D·S '99

A Note on Prices

It is impossible to make accurate comparisons between prices in the Golden Age of the Dutch Republic and those today. Figures can certainly be calculated, based on the comparative prices of gold or essential foodstuffs, but they do not take into account vital differences such as what constitutes a minimal standard of living (in many respects, people who today would be called poor live more comfortably than the richest Dutch in the seventeenth century), and certainly not what luxuries such as tulip bulbs were worth in the Golden Age.

The best comparisons probably come from looking at different salaries and earnings. The table below sets out some typical examples from the Dutch Republic in the first half of the seventeenth century.

The basic unit of currency in the Republic was the guilder. One guilder was made up of 20 stuivers.

½ stuiver	Cost of a tankard of beer
6½ stuivers	Cost of a 12lb loaf, 1620
8 stuivers	Daily wage of an experienced Haarlem bleacher, 1601 (=about 110 guilders a year)
18 stuivers	Daily wage of an Amsterdam cloth-shearer, 1633 (=about 250 guilders a year)
13 guilders	Exchange price of one Dutch ton of herring, 1636
60 guilders	Exchange price of 40 gallons of French brandy, 1636
250 guilders	Annual earnings of a carpenter, 1630s
750 guilders	Clusius's salary at the University of Leiden, 1592
1500 guilders	Typical earnings of a middle-ranking merchant, 1630s
1600 guilders	Rembrandt's fee for his greatest masterpiece, *The Night Watch*, 1642
3000 guilders	Typical earnings of a well-off merchant, 1630s
5200 guilders	Highest reliably attested price paid for a tulip bulb, 1637

Sources: van Deursen, Hunger, Posthumus *Inquiry*, Zumthor

Introduction

One day in the early spring of 1637, a Dutch merchant named François Koster paid the enormous sum of 6650 guilders for a few dozen tulip bulbs.

At a time when a whole family might live for a year on perhaps 300 guilders, this was a remarkable purchase. Even more surprisingly, Koster had absolutely no intention of actually growing his tulips. He planned to take the bulbs and sell them on, and fully expected to make a profit from the deal.

There were many, even then, who thought Koster and the thousands of other Dutch people who fought for the right to buy ordinary flower bulbs at incredible prices were insane, but in truth the tulip traders had every reason to suppose that the prices they had paid were justified. The most valuable bulbs were rare, and the flowers they produced were greatly coveted and extraordinarily beautiful. Tulip prices had already risen swiftly and consistently for more than two years. Why shouldn't they go higher still?

Nevertheless, the traders were wrong. Koster's extravagant

purchase took place at the height of one of the most bizarre and memorable episodes in history, the great Dutch tulip mania of 1633–7. It proved to be one of the last expressions of the mania at work. Less than a week after the merchant had bought the bulbs, tulip prices dropped, suddenly and without warning. Within a matter of days, flowers had plummeted to only one-tenth of their old values, and often considerably less. By the end of February 1637, men who had been – at least on paper – among the richest citizens in all Holland had lost everything they had, and those who had invested heavily in tulips faced bankruptcy and ruin. François Koster himself, having made a down payment of 820 guilders for his bulbs, found himself unable to pay the balance of the purchase price, 5830 guilders, and was pursued remorselessly through the courts by the irate dealer who had sold him the flowers.

The story of the flower mania so completely contradicts what we are generally taught about the history of the seventeenth century that it begs for explanation. How could people have become so distracted by something so apparently superficial in an era full of war and want? How could a society which held success to be synonymous with virtue, which publicly professed the strictest form of Calvinism, banned the use of fripperies such as church organs and even frowned on dancing at wedding feasts, tolerate the greed and extravagancies of a trade chiefly carried on by drunken men sitting in the back rooms of taverns? And what was it, anyway, that made those men deal in flower bulbs and not some other more conventional form of merchandise?

Even at the time, there was a feeling that something quite

remarkable had happened. Beginning when the mania was at its peak and continuing through the collapse in prices, the Dutch themselves produced a deluge of pamphlets which sought to explain or satirize the flower trade. Those who lived outside the borders of the country (which was then known as the United Provinces of the Netherlands) looked on with even greater incredulity as a people famous throughout Europe for being dour, drab, sternly moralistic and above all extraordinarily canny in matters of finance abandoned themselves, it seemed completely, to an inexplicable passion for tulips.

The people who lived through the bulb craze found it difficult to explain why it occurred, and in most respects we are not that much wiser now. Today, more than 360 years later, there is still no proper history of the subject based on archival sources. The scant original work which has been done was completed as long ago as 1934, and is in any case only available in Dutch. Since then, the few authors who have made more than just a passing reference to tulip mania have mostly copied their information from earlier works which are not always very reliable. And yet the tulip fever is such an inherently memorable subject – once encountered, never forgotten – that many assume it must have been thoroughly explored.

Over the years thousands of general readers have been introduced to tulipomania by the work of a nineteenth-century Scottish journalist called Charles Mackay, whose book *Extraordinary Popular Delusions and the Madness of Crowds* – first published in 1841 and still in print today – contains the classic English-language account of the craze. Indeed, the weird hold that the tulip mania exercises on the imagination is well illustrated by the fact that although Mackay's book is 725 pages long and contains sixteen chapters on subjects ranging from the history of the Crusades to 'The influence of politics and religion on the hair and beard', it is for his tiny eight-page section on tulips that he is generally remembered. Unfortunately, while Mackay's opinions have been influential, many of his facts are misleading or plain wrong. Few accounts of the mania have not been coloured by his mistakes, and when the story is told, it is, thanks to Mackay, almost always recounted in tones of bemused incredulity. The country was crazy, the dealers mad, the stories go. It cannot really be explained.

Economists and city traders, too, are superficially familiar with tulipomania. 'Gathered around the campfire early in their training,' as one professor writes, 'fledgling economists hear the legend of the Dutch tulip speculation from their elders, priming them with a sceptical attitude towards speculative markets. That the prices of bulbs could rise so high and collapse so rapidly seems to prove a decisive example of the instability and irrationality that may materialize in asset markets.' But financiers and economists have built their analyses of the mania on the same dubious foundations as the historians. (It is even said that some of the more literate investment banks on Wall Street still hand their new employees copies of Mackay's book and tell them to absorb his chapters on the causes of financial catastrophes before they are let loose on a trading floor.) For them, tulipomania is the first great mania, the precursor to financial crises of truly seismic proportions, such as the South Sea Bubble (which saw some of the most notable members of London society lose thousands of pounds in a fraudulent scheme for trading with South America). It has been allotted a neat position as the very first in a long succession of booms and busts which begins with the introduction of paper money, stocks and shares, and ends – for the time being – with the great crash of 1989. Whenever stocks or bonds appear to be valued at more than they are worth (for instance, shares in companies that try to make money from the Internet), comparisons are drawn in the financial press with tulipomania. Because economists depend not on interpretation and impressions but on facts and figures – which are hard to come by in the records of the tulip craze – it is probably more difficult for them to see the story in its proper context than it is for the historians. Even today there are still fundamental differences between professors who see the tulip mania as a classic example of a bubble – that is, a sharp rise in the price of something of no real worth – and those who believe that the huge sums the bulbs commanded were justified by the fact they were in high demand and short supply.

Tulip mania happened to people whose attitudes were recognizably modern in many ways, yet who in other respects led quite different lives from us and from the people who live in the Netherlands today. The characteristics which are now widely associated with the Dutch – tolerance, stolidness, love of beer, a certain multiculturalism – were already being attributed to Hollanders in the seventeenth century. And it is worth remembering that the trading frenzy coincided with the height of the Dutch Golden Age, a brief period during which the United Provinces enjoyed not just worldwide commercial supremacy (Amsterdam was to the seventeenth century what London was to the nineteenth) but also an astonishing cultural richness. Rembrandt and Vermeer both lived through the tulip mania, and many of the greatest of all Dutch artists painted tulips.

The tulips of the time, moreover, were absolutely beautiful, much more so than the relatively plain varieties available today. The most valuable flowers of this period possessed petals adorned with vivid and intricately patterned flames of colour which had never been seen before and have since disappeared for ever. They were irresistible.

This, then, is the true story of the tulip mania. It is the story of how the flower came to leave its original homeland in the East and travel thousands of miles to the United Provinces, how it established itself there, and why the bulb craze occurred where and when it did. It is an attempt to set the history of the craze straight and to understand the financial realities which underpinned it.

It is also the story of two different tulip manias, for events in the Dutch Republic were mirrored by an equally remarkable craze which took place among the Turks of the Ottoman Empire. There, the tulip was regarded not as an object of speculation but as a holy flower. It was grown in secret gardens at the heart of the most sacred of imperial palaces, and enjoyed the favour of several of the all-powerful sultans who terrorized half of Europe with their armies – men who thought nothing of ordering a hundred executions with one breath and 10,000 bulbs from Syria with the next. The contrast between the tulip's experiences in Holland, where it was the subject of a financial mania, and Turkey, where the passion for it was of a more cultural and artistic nature, is as fascinating as it is instructive.

So little of real usefulness has been written on the subject of the tulip craze that it is doubly unfortunate that much of what has been said is wrong. Did everybody living in the United Provinces, from the richest merchant to the poorest vagabond, actually find themselves caught up in the mania? Were sums the equivalent of almost a million pounds paid for single bulbs? When the crash in flower prices came, was it really so severe that it threw the whole Dutch economy – which was then the richest and fastest-growing in the world – into recession? Did an all-powerful Turkish sultan truly lose his throne simply because he, too, became obsessed with tulips to a dangerous degree?

The answer to all these questions is no. But that need not diminish the interest of the tulip mania. It actually did occur, some people did make fortunes, others truly were ruined; and anything capable of making weavers richer than spice merchants, and poor orphan children so wealthy they could contemplate never having to work a day in their whole lives – as the Dutch flower craze really did, for however brief a time – possesses a genuine fascination quite independent of the myths which have accumulated around it.

PROLOGUE

A Mania for Tulips

They came from all over Holland, dressed like crows in black from head to foot and journeying along frozen tracks rendered treacherous by the scars of a thousand hooves and narrow wheels. They had cloaked and blanketed themselves against the biting winter wind – the wealthiest rattling along in unsprung carriages that jerked from rut to pothole like an untried sailor lurching through a hurricane, the rest on horseback with their heads bowed against the cold. Travelling singly or in twos and threes, they clattered through the flat and sterile landscape north of Amsterdam, riding on until they came to the little town of Alkmaar near the coast.

They were middle aged and stoutly built: shrewd and successful men who had made their money in trade, who knew how to turn a profit and what it meant to live well. Most were clean shaven and ruddy-faced; their clothes, though drab, were cut from the finest cloth, and the purses that they carried were snugly full of money. Passing through the gates of the town at dusk, the visitors made their way through Alkmaar's cramped and narrow streets and found rooms in taverns near the busy market place. There they ate and drank and puffed their long clay pipes into the night, calling for great pitchers of wine and plates of roasted meats, sprawling back in their hard wood chairs and talking shop till past midnight by the jaundice-yellow light of the peat fires in the grates.

The business of these rich Dutch merchants was not grain or spices, timber or fish. They dealt, rather, in tulip bulbs – drab and anonymous brown packages which resembled nothing so much as onions. Yet, unpromising as they might at first appear, flowers at this time were far more precious than the richest commodities that could be found piled up on the wharves of Amsterdam. Some tulips were so scarce and so greatly coveted that they were worth more than a hundred times their weight in gold, and successful bulb dealers could make huge profits. In an age when the richest man in the whole of the United Provinces was worth 400,000 guilders – a sum amassed over several generations – some tulip traders were buying and selling single flowers for hundreds, even thousands, of guilders, and building paper fortunes of as much as 40,000 or 60,000 guilders in a matter of a year or two.

The bulb dealers had come to Alkmaar to attend an unprecedented auction. The guardians of the little orphanage in the town had come into the possession of one of the most valuable collections of tulips in the whole of the Netherlands. Caring more for the flowers' value than their beauty, they were selling off the bulbs for the benefit of some of the children in their care. So, shortly after dawn broke, grey and chill, the traders began to make their way to the saleroom in the Nieuwe Schutters-Doelen – the headquarters of Alkmaar's civic guard – an ornate and gabled building in the centre of the town.

It was a large room, but they filled it, a pressing mass of sombre burghers in the throes of avarice. When the auctioneer appeared, the bidding started briskly and soon became frantic. Single bulbs were knocked down for 200 guilders, then 400, 600, a thousand and more. Four of the hundred or so lots were sold for in excess of 2000 guilders apiece. And when at last the final tulip had been sold and all the money tallied, the auction proved to have raised a total of 90,000 guilders, which was, quite literally, a fortune in those days.

The date was 5 February 1637, the day flower fever reached such a pitch of frenzy in the United Provinces that once worthless bulbs truly threatened to supplant precious metals as objects of desire. The day the tulip completed a journey which had begun hundreds of years before and thousands of miles away.

CHAPTER I

The Valleys of Tien-shan

The tulip is not a native of the Netherlands. It is a flower of the East, a child of the unimaginable vastness of central Asia. So far as anyone can tell, it did not reach the United Provinces until 1570, and by then it had already been journeying for many hundreds of years from its original homeland in the mountain ranges that run north of the Himalayas along the fortieth parallel.

Taxonomists believe that the first tulips sprang from the scrubby slopes of the Pamirs and flourished among the foothills and valleys of the Tien-shan mountains, where China and Tibet meet Russia and Afghanistan in one of the least hospitable environments on earth. These flowers were relatively sober and compact things, with narrower petals than Dutch tulips. The tulips of the Tien-shan were much shorter than modern tulips, carrying their flowers usually a scant few inches above the ground, but they were hardy and well adapted to the harsh winters and parched summers of central Asia. They were predominantly red, the colour of blood or soldiers' uniforms, and they were venerated by the warlike tribes who peopled this desolate area. Yet nothing could have been less regimented, less militaristic, than the scattered colonies of scarlet tulips that clung to the barren soil of these mountain ranges. They were not uniform but infinitely varied, each flower differing subtly from its neighbours in its colour or the shape of its petals; and they were not conquerors but seducers, flowers which wormed their way into a man's soul.

These tulips were not the finished article – not yet. They lacked the striking colour schemes that distinguished the flowers that were to entrance the Ottoman Empire and cause Dutchmen to abandon both their caution and their common sense, the contrasting streaks and flares of pigment that made each bloom a living canvas. They had neither the stature nor easy elegance that characterized their descendants. These would come only with time. But, even now, they were beautiful.

Nearly half of the 120 known species of tulip grow wild in this forbidding terrain. Together, the Pamirs (Russia's 'roof of the world') and the Tien-shan – the 'celestial mountains' which run along China's western border – form the backbone of Asia, an all but impenetrable barrier several thousand miles long and hundreds of miles wide. Thousands of years ago, these mountains were the reason why the ancient civilizations of Rome and China remained almost entirely ignorant of each other's existence; today they remain among the least explored regions on earth. As late as 1900, when Britain had occupied India and Russia had subdued the fastness of Siberia, this inner Asian citadel remained unexplored by Europeans. Bordered to the east by impassable, bone-dry desert, to the north by barren taiga, to the west by warring, hostile khanates and to the south by mysterious and unwelcoming Tibet, the craggy fortress of the Tien-shan was as inaccessible as any place on earth. Even the valleys of this immense range were found at such altitudes that the few outsiders who visited them had to acclimatize to the lung-searing mountain air, and the passes which led to more hospitable country could not be crossed for eight or nine months of each year. When, at the height of summer, the worst of the snows did melt, the Tien-shan remained impassable to all but the hardiest travellers, a sea of gneiss and granite which contained no settlements, no soil worth cultivating, and little or no water. They are dry, infertile and unwelcoming – incapable of supporting either plant or animal life.

Yet even the Celestial Mountains and the Roof of the World boast occasional oases and foothills where life can flourish. In the case of the Tien-shan, the valleys lie predominantly on the north side of the range, with the oases and settlements and the trade that they attract along the foothills to the south. These towns were a considerable lure for the Turkish nomads who have peopled the Asian steppe lands since the beginning of recorded history. Pasturing their horses in summer in the rich valleys of the north, and crossing the mountains through little-used passes, they would descend occasionally on the cities of the south – sometimes pillaging and raiding, sometimes trading with the civilizations of the oases for their learning and their silk.

As pastoralists, the Turks would have encountered the tulip where it grew wild in the valleys of Tien-shan; as invaders, they would also have found colonies growing at much higher altitudes as they crossed the passes leading south, for the tulip can flourish in very mountainous terrain and even winter under a blanket of snow. The simple beauty of these unsophisticated wild flowers, with their petals coloured yellow or orange or cinnabar, must have been considerably enhanced by the bleak surroundings in which they were usually encountered. For nomads who had survived another howling, freezing Asian winter, the year's first tulips were more than just patches of beauty appearing in the wilderness. They represented life and fertility. They were the heralds of spring.

Tulips, then, became an important symbol for the Turks. As they moved westwards across the endless steppe, the nomads found colonies of the flower growing wild all across the central Asian plateau, from Tien-shan to the Caspian Sea, and then along the further reaches of the Black Sea coast and south among the Caucasus. These tulips had spread westwards naturally, thousands of years earlier. But by the time migrating Turks appeared in numbers in the Middle East, in the tenth and eleventh centuries AD, some at least of the flowers that they encountered were growing in gardens, planted where they might best please the eye.

When exactly cultivation of these wild flowers began is a mystery, but we do know that by about the year 1050, tulips were already venerated in Persia. Tulips grew in the gardens of the old Persian capital, Isfahan, and also in Baghdad. They appear in one of Omar Khayyám's best-known verses as a metaphor for perfect female beauty, and later poets often used the flower as a symbol of perfection. One, Musli Addin Sa'adi, described his ideal garden in about 1250 as a place where 'the murmur of a cool stream, bird song, ripe fruit in plenty, bright multi-coloured tulips and fragrant roses' combined to create an earthly paradise. Another, Hafiz, likened the sheen of the flower's petals to the bloom on his mistress's cheek.

Indeed, the tulip's delicacy and typically blood-red colouring made it a flower of great symbolic importance for the people of Persia. It was synonymous with eternity, and several myths were told to explain its outstanding beauty. One such legend told how a prince named Farhad was deeply in love with a maiden, Shirin. One day word reached him (falsely, as it turned out) that his beloved had been killed. Gripped by unbearable grief, he hacked his own body open with an axe. Blood dripping from his terrible wounds fell on the barren soil, and from each drop a scarlet flower sprang, a symbol of his perfect love. Hundreds of years after this story was first written down, wild red tulips remained a favourite Persian token of undying passion. 'When a young man presents one to his mistress,' the seventeenth-century traveller John Chardin recounted, 'he gives her to understand, by the general colour of the flower, that he is on fire with her beauty; and by the black base of it, that his heart is burnt to a coal.'

Among the largely illiterate Turkish peoples of the steppe, no records exist which trace the flower's history further back than Omar Khayyám's day, and it is not until the end of the eleventh century, when a tribe of Turks called the Seljuks came west and seized Anatolia from the Byzantines, that the tulip first appears in nomad art. The Seljuks either brought the flower with them as they began to explore the land, or discovered colonies of wild flowers where they settled. The earliest known drawings of tulips are found on tiles excavated from the thirteenth-century palace that one of their sultans, Alaeddin Kaikubad I, built on Lake Beysehir in eastern Anatolia.

By this time, the Turks had lost some of their nomadic instincts. The Seljuks settled in the cities that they captured and they called the lands they had taken 'Rum' because they saw themselves as the inheritors of Rome. They certainly developed a Roman taste for empire-building, and even after the sultanate of Rum was annihilated at the beginning of the fourteenth century, Seljuk princelings began to carve new kingdoms from its ruins.

One of these petty rulers was a certain Osman of Sogut, and his dynasty (known as 'Othman' to the Arabs and as 'Ottoman' in Europe) proved to be the most glorious in all the long history of the Turks. It was a house of conquerors and despots which enslaved great swathes of Asia and swept through Europe to the gates of Vienna – a line whose rulers not only held the power of life and death over their subjects, but frequently used it. Yet many of the Ottoman rulers were also cultivated men with delicate tastes and a passion for beauty. They dressed in the most sumptuous cottons and silks, ate songbirds and rewarded expert calligraphers as richly as successful generals. The Ottomans were also knowledgeable horticulturalists, and eventually they elevated the tulip to a position of eminence it had never enjoyed before.

By 1345, the House of Osman had crossed the Dardanelles and Turkish horsemen had arrived in Europe. They came at the request of the emperor of Byzantium, who wanted their help to secure his throne against a usurper. Instead, the Ottomans took Greece and Thrace for themselves, and then most of the Balkans as well, reducing the emperor to a puppet whose writ seldom ran more than a few miles beyond the walls of his great capital, Constantinople.

It is impossible to be sure how widespread the cult of the tulip was among the Ottomans who swarmed across the Balkans in the first half of the fifteenth century. The Turks of this era generally obeyed Islam's proscriptions against the public display of realistic portraits of living things*, and because of this, there are no representations of the tulip in Ottoman manuscripts of the period – no contemporary paintings, vases decorated with flowers, or tulip-emblazoned tiles appear to have survived, if any were ever made. Nevertheless, we know they venerated gardens and regarded flowers as sacred.

The Turks told a story to explain why gardens were so important to them. When Hasan Efendi, a famous dervish holy man, was preaching one day, one of those who had come to hear him speak passed him a note. It enquired whether any Muslim could be certain he would go to paradise when he died. After Hasan had finished his sermon, he asked if there were any gardeners present. When one member of the congregation stood up, Hasan pointed to him and said: 'This man will go to heaven.'

Immediately the dervish was surrounded by a press of people demanding to know what the gardener had done to be certain of a place in paradise. But Sheikh Hasan explained that he was merely quoting from the *Hadith* – the oral traditions of the Prophet Muhammad – which state that people will do in the afterlife what they most enjoy doing

* The reason for this was that it was thought insulting for man to attempt to capture – imperfectly – one of the perfect creations of God.

on earth. Because all flowers belong to heaven, gardeners will surely go to paradise to continue their work.

Indeed, the garden is central to the Muslim vision of paradise. Christian clerics told their flocks that heaven was a shining city on a hill; the Arab founders of Islam, a religion which had after all sprung originally from the desert, looked forward to an endless garden of delight, full of pavilions and fountains, carpeted with flowers of a beauty unequalled on earth. Pious Muslims treated flowers almost as holy relics and often wore blooms in their turbans.

The tulips of the Persians and the Turks were still wild flowers. Even when they were planted in gardens, they were not yet cultivated in the sense of being systematically bred, crossed with other strains or otherwise improved by human intervention. As late as the early sixteenth century, when the Turkish warlord Babur counted thirty-three different varieties of wild tulip as he passed south through Afghanistan, the old nomad peoples do not seem to have encountered any garden hybrids. When Babur – who overthrew the kingdoms of northern India and established the dynasty of Moguls, whose name remains a byword for luxury and opulence – planted tulips in the innumerable formal gardens he created, the bulbs he sowed were wild-flower bulbs.

Of all the blooms in a Muslim garden, the tulip was regarded as the holiest, and the Turkish passion for this flower went far beyond mere appreciation of its beauty. For the Ottomans, as for the Persians, it had a tremendous symbolic importance and was literally regarded as the flower of God because, in Arabic script, the letters which make up *lale*, the Turkish word for 'tulip', are the same as those which form 'Allah'. The tulip also represented the virtue of

modesty before God: when in full bloom, it bows its head.
After the proscription on images of living things was finally
relaxed, in the course of the fifteenth and early sixteenth
centuries, tulips were often depicted in Ottoman illustrations
of the Garden of Eden, blossoming beneath the fruit trees
where Eve was tempted. Turks who willingly gave their lives
in battle, believing death in the service of Islam was the
surest passport to a paradise of meadowlands where divinely
beautiful *houris* would serve them the wine they were denied
on earth, fully expected to find their heaven strewn with
tulips. To an Ottoman gardener, therefore, it was one of
the most precious of flowers – only the rose, the narcissus,
the carnation and the hyacinth were worthy to be planted
alongside it. All other blooms, however rare, however
beautiful, were considered 'wild flowers', and were cultivated
only occasionally. For this reason, it is easy to believe that
tulips accompanied the Turks as they swept westwards from
Asia into Europe.

CHAPTER 2

Within the Abode of Bliss

Two hundred and fifty years before Dutchmen bid for bulbs in the auction-rooms of Holland, the tulip came to the plain of Kosovo in the southern marches of Serbia. There, at a place called the Field of Blackbirds, a Christian army of 15,000 men led by a man named Prince Lazar stood and faced twice that number of Ottoman Turks under the command of their sultan, Murad I. The great battle which Murad and Lazar fought on St Vitus's Day in 1389 helped to seal the fate of the Balkans for the next five hundred years.

The day did not begin well for the Serbs. The best and bravest Christian knights were beaten back, and Lazar himself was captured in the confusion. On the Turkish side, meanwhile, Murad directed his men with the skill to be expected of a sultan who had spent most of his thirty-year reign on campaign. His position at the centre of the Ottoman army seemed secure; he was screened by three lines of camels, chained one to another to present an impenetrable obstacle to the Christian cavalry and intended, like Hannibal's elephants, to terrify an enemy which had never encountered such exotic creatures before. And yet, somehow, one Christian soldier did reach the sultan. According to legend, this man was a Serb whom Lazar had publicly accused of treachery on the previous evening, and who now proved his loyalty by impaling Murad with such force that the dagger thrust into the Turk's chest sprouted from his back.

The sultan fell, mortally wounded, but he remained alive just long enough to summon the captive Prince Lazar and order his immediate execution. Thus the leaders of the Christians and of the Turks joined the thousands of their men who lay dead upon the Field of Blackbirds – so many that a Muslim chronicler recalling a battleground thickly covered with the fallen, and strewn with severed heads still wearing brightly dyed turbans, wrote that he was put in mind of a gigantic bed of tulips, their gaudy red and yellow petals echoing the brilliant colours of the Turkish headdresses.

In fact, it is possible that tulips really were represented at the Battle of Kosovo – not merely by the Turkish headdresses, in the poetic phrase of the chronicler, but in the form of talismans. In the fourteenth century, the Ottomans seem to have adopted this most holy of flowers to guard them against misfortune, albeit in a slightly peculiar way. Partly for protection, and partly because the religious proscription against images of living things still had force, the tulip was embroidered not on banners and surcoats, but on underclothes. The Museum of Turkish and Islamic Arts in Istanbul still displays a simple cotton shirt, made to be worn beneath armour and richly decorated with verses from the Koran on the front and embroidered tulips on the back, which was taken from the tomb of one of the Ottoman generals who fought at Kosovo. This general was Sultan Murad's second son, Bayezid, a young prince who had scarcely reached manhood when he led a division of the Turkish army against Prince Lazar. Bayezid is the first man in history who can be personally identified with the tulip.

He is supposed to have donned the shirt as a protection against evil, but also as a good-luck charm. If that is so, the flower served him well at Kosovo. Proclaimed sultan by his men, Murad's younger son succeeded his father on the Field of Blackbirds while the battle against the Serbs still raged. He began his reign – quite ruthlessly – by ordering the execution of Yakub, his elder brother and chief rival for the throne. This unfortunate prince was quickly garrotted with a silken bowstring in compliance with Bayezid's decree. The new sultan thus secured the Ottoman succession for himself under the most testing of circumstances.

Bayezid proved to be a ruler of immense energy and ambition. He tightened the Ottomans' grip on the Balkans and, in 1396, utterly defeated the last great crusading army, a force of some 16,000 men, at Nicopolis in Bulgaria. After the battle, the sultan personally supervised the beheading of about 3000 Christian captives. It was hardly surprising that his subjects began to call him *Yildirim*, 'the Thunderbolt'.

For fully thirteen years, in fact, Bayezid triumphed at every turn, crushing Christian resistance in the Balkans and slaughtering Persians in the east. But the power of his talisman had now exhausted itself. In 1402, near Ankara, he fought a ruler even greater and more implacable than himself: Tamerlane, a crippled Mongol born in the shadow of the Pamirs, a soldier almost as able as his predecessor,

Genghis Khan, but even more bloodthirsty. Bayezid's army was scattered, and the sultan himself was overtaken by Mongol archers as he fled the field and brought to grovel at the feet of his conqueror in Tamerlane's own tent.

The tulip king was shown no mercy. Tamerlane seized the women of the sultan's harem for himself, and forced Bayezid's wife Despina to wait on him, naked, at his table. The sultan he confined within an iron cage, which the Mongols took with them as they travelled. On state occasions, Tamerlane had the once proud Bayezid dragged before him so he could use him as a footstool.

Bayezid survived only eight months of this treatment. His end remains obscure; some say he died of apoplexy, but the playwright Christopher Marlowe, in *Tamburlaine the Great*, has him dash out his own brains against the bars of the cage in despair at his plight. At any rate, he was dead before the tulips flowered in 1403.

The sultan's capture temporarily halted the tulip's westward progress and left the fledgling Ottoman Empire in a state of chaos from which it took half a century to recover. The principal beneficiaries were the shattered remnants of the Christian states which had ruled the Balkans before the sultan's time, and particularly the Greeks of Byzantium. Bayezid's greatest ambition had been to take Constantinople and make it the new centre of his empire, and he had even besieged the city for five years at the end of the fourteenth century, but he was never able to break down the massive fortifications that enclosed it.

Constantinople was something of a shadow city by 1400, its decline reflecting the fading fortunes of its Byzantine rulers. In fact it was more than half-empty, the 7 long miles of its walls enclosing a town of no more than 50,000 people, scattered now among what were effectively large villages separated by ruins, working farms and orchards. But in size and situation and repute it was still the greatest city in the world. It was fit to be the capital of the Ottoman Empire – and the new home of the tulip.

Bayezid's demise did not save the Byzantines; it merely postponed their end. Within half a century, the Ottomans had regrouped and returned under the command of the dead sultan's great-grandson, Sultan Mehmed. This time Constantinople was weaker, and the Turkish army considerably larger and equipped with the latest cannon and catapults. After a desperate siege lasting less than two months, Mehmed's troops forced a breach in the walls and the Turks poured into Constantinople. The last Byzantine emperor threw away his imperial insignia and sought an anonymous death in the press of the fighting. Then, amid terrible scenes of massacre, the Ottomans took Constantinople and made it Istanbul.

Even by the remarkable standards of the Ottoman sultans, Mehmed – who was henceforth always known as Mehmed the Conqueror – was a complicated character. Warlike but cultured, sensuous but implacable, he was a ruthless monarch but a humble man. When he gave thanks for his victory at the Byzantine cathedral of St Sophia on the day Constantinople fell, he knelt and scattered a handful of earth over his turban as an act of obeisance to God. He was also the author of a gloomy Turkish couplet:

> Footman, pour me some wine, for one day
> the tulip garden will be destroyed;
> Autumn will come soon, and the spring season
> will be no more.

Mehmed may have been a realist, but he had no intention of relinquishing the Ottomans' hold on their new capital just yet. On the contrary, the once great city began to recover under his rule. New buildings appeared on the skyline; four huge minarets rose alongside St Sophia, which became the Aga Sofia mosque; the land walls were repaired, new palaces begun. And in places that had been abandoned to ruin under Byzantine rule, the Turks built a myriad of gardens.

Blessed though it was by one of the most perfect physical situations in the world, Istanbul craved such adornment. It had been built on seven great hills at the very edge of Europe, with water on three sides. Even as the Byzantines had left it, the city offered gorgeous views at every turn. Taking full advantage of its emptiness, the Turks planted trees and flowers so their natural beauty complemented and offset the city's buildings, old and new. Within a few decades of the conquest, the Ottoman sultan alone could enjoy more than sixty private gardens scattered along the Bosphorus and the Sea of Marmara. Dozens more kitchen plots supplied fruit and vegetables to his palaces. Other Ottomans built sunken gardens which offered shade in the heat of the summer, terraced gardens full of vines, pleasure gardens in public places, and private 'paradise gardens' enclosed within the walls of their own homes and filled with flowers.

This profusion of greenery distinguished Istanbul, in the eyes of visitors, from any European city. And its inhabitants

planted their gardens in ways that startled Western horticulturalists. The Turks hated the corseted regimentation of
the formal gardens that were in fashion at the courts of
England, France and Italy. Ottoman gardens were impres-
sionist spectaculars in comparison, planted not to impress
the eye with geometrical precision, but to seduce it with
visions of lushness and of plenty. An Ottoman garden was
designed as a place where its owner might seek refuge from
the cares of the world and a retreat from the heat of the
day. Within its walls the Turks grew soft fruits and created
fountains and melodic streams. It was intended as a little
piece of heaven here on earth.

Europeans who travelled to Istanbul during the high days
of the Ottoman Empire that Mehmed and his successors
now built were generally surprised not merely by the city's
size and opulence, but by its masters' manners and good
taste. This was a city of culture and coffee-houses, which
tolerated the religious diversity of its inhabitants in a manner
inconceivable in Europe. Yet the Western notion of the
Turk was all to do with cruelty and lust – the savagery of
the Ottoman armies was a popular theme, as was curiosity
about the hidden pleasures of the sultan's harem – and
certainly the Turks themselves were as capable of cruelty as
they were appreciative of beauty.

Sultan Mehmed himself was a man of just such con-
tradictions. One of his earliest acts was to order the construction of a wonderful new palace at the eastern end of
the city, poetically named the Abode of Bliss by its creator
but better known today as the Topkapi. It was specifically
intended to outdo in its magnificence anything built during
the Byzantine millennium, combining – in the words of one

chronicler – 'variety, beauty and magnificence'. In it 'on every side, inside and out, shone and glittered gold and silver, ornaments of precious stones, and pearls in abundance'. Mehmed, a passionate gardener who collected rare plants from every part of his domain and could often be seen labouring in person among his flowers, saw to it that the Abode of Bliss was surrounded by 'very vast and very beautiful gardens, in which grew every imaginable kind of plants and fruits; where water, fresh, clear and drinkable, flowed in abundance on every side, and flocks of birds, both of the edible and of the singing variety, chattered and warbled'. Yet when this cultured man discovered one day that one of his prized cucumbers had been stolen, he had the palace gardeners brought before him and disembowelled, one by one, in the hope of ascertaining which of them had eaten it.

Later Ottoman rulers more than matched Mehmed the Conqueror in both cruelty and their enthusiasm for exquisite palaces and gardens. The greatest of them all was Mehmed's great-grandson Suleyman the Magnificent, who came to the throne in 1520 and stretched the Turkish empire from the gates of Vienna to the Persian Gulf and from the Straits of Gibraltar to the Caspian Sea. To Europeans Suleyman was 'the Grand Turk', the title by which subsequent sultans were also known to the West, and he was acclaimed, among his many other titles, 'Possessor of Men's Necks'. He was a byword for ruthlessness among those Christians unfortunate enough to encounter his armies. But Suleyman's subjects revered him as 'the lawgiver', and he was a pious man who – exceptionally for an Ottoman – had little use for the harem and lived a chaste life with his favourite wife.

By Suleyman's day, in the first half of the sixteenth century, the tulip had established itself as the quintessential Turkish flower. It was still unknown in Europe, but its popularity among the sultan and his servants was such that – now the old proscription on the portrayal of living things was being relaxed – it had become one of the favourite motifs of Ottoman artists and artisans, appearing with increasing frequency on flower vases and tiles. Tulips graced the sultan's robes, and not merely his underclothes as they had done in Bayezid's time: Suleyman's cream-coloured imperial brocade gown, which still survives, was embroidered with hundreds of blooms. The royal armour, worn on campaign in Hungary and Persia, was embossed with a single glorious tulip, 9 inches long, and the sultan's helmet, a masterpiece of the armourer's craft, was adorned with tulips shaped in gold and set with precious stones.

By the middle of the sixteenth century, tulips were becoming much more commonplace within the Ottoman Empire, and other Turks beside the sultan were also making copious use of the flower. They were embroidered on the prayer rugs sewn by brides for their trousseaux and painted on water bottles or woven into the velvet coverings that ornamented elaborate Turkish saddles. And just as gardeners planted tulip bulbs to help their souls ascend to paradise, so the women of the Turkish empire sewed thousand upon thousand images of the flower as religious tokens, and offered them up with prayers for a husband's safe return from war.

It was under Suleyman, it seems, that the Turks first began to cultivate the tulip and to breed new varieties to suit their tastes. The wild flowers which had been grown in Istanbul since Mehmed's day were short and rounded, almost egg-shaped, and not unlike many of the varieties still popular today. Perhaps as early as the late sixteenth century, however, the Ottomans began to look with favour on new cultivars* which the capital's gardeners had begun to produce. These 'Istanbul tulips', as they became known, may have been bred from species which the Turks had discovered on the northern shores of the Black Sea, in the land of their allies the Crimean Tatars. Istanbul tulips – of which there were eventually as many as 1500 varieties – were more delicate and elegant than their predecessors. Their petals were enormously long and slender, and needle-pointed at the tip. The most sought-after varieties were shaped like almonds, with daggerlike tepals. They were coloured vermilion or russet or sulphur.

The first gardeners to devote themselves entirely to tulips lived in Suleyman's time and grew some of the earliest cultivated tulips. One, named Seyhulislam Ebusuud Efendi, is known to have possessed a particularly beautiful flower known as Nur-i-Adin, 'The Light of Paradise'. Other varieties of the flower were given equally evocative titles which reflected their value and their beauty: Dur-i-Yekta, 'The Matchless Pearl'; Halet-efza, 'Increaser of Pleasure; 'Instiller of Passion'; 'Diamond's Envy'; 'Rose of the Dawn'.

To begin with, such tulips were great rarities. Even Seyhulislam – who died, at the greatly advanced age of eighty-four, in 1574 – would have possessed only a handful of bulbs of the Nur-i-Adin. And in an age when the art of

* Flowers systematically cultivated and improved.

coaxing new varieties from old was barely understood, so that growers who wished to produce crimson flowers might attempt to do so by pouring dark-red wine over their tulip beds, cultivation was a slow and somewhat haphazard business, and one which failed to interest most Turkish gardeners. The majority of new Ottoman cultivars seem to have emerged by accident rather than design.

Nevertheless, the Ottoman sultans gradually increased their stock of bulbs, and used tulips and other flowers to adorn their palaces and gardens. Some of these blooms were grown in Istanbul, where there were, by the 1630s, about eighty flower shops and three hundred professional florists. Others were imported, sometimes in great bulk. New varieties of tulip came from the Black Sea coast and Crete, or from Persia, taken by force during one of the interminable campaigns the Ottomans fought there. In 1574 Suleyman's son Selim II – a keen gardener whose other passion, alcohol, led to his becoming known to history as Selim the Sot – instructed the Sheriff of Aziz, in the Turkish province of Syria, to send him 50,000 tulip bulbs for the imperial gardens. 'I command you not in any way to delay,' the sultan added. 'Everything should be so well and quickly done that it should give rise to no disappointment.' Even though Selim made it clear that money to pay for the purchases could be had from the treasury in nearby Aleppo, such orders must have caused great consternation in those receiving them, as perhaps the sultan intended.

Of all the Grand Turk's gardens, those hidden within the walls of his own home, the Topkapi Palace, were by far the most opulent. But then everything about the Abode of Bliss was meant to demonstrate the magnificence, wealth and taste of the Ottoman royal line. Even the public portions of the palace were built on the grandest scale, and the private quarters, which only the highest-ranking Turks and their personal servants usually saw, were of a size and complexity unrivalled in the West.

In order to reach the inner sanctums where the sultan's tulips were displayed, a visitor would have had to approach the Abode of Bliss via a thoroughfare which led past the Aga Sofia mosque and opened on to a plaza. Once there, he would have seen the palace's outer walls, bristling with fortifications and guards and pierced by a huge outer gate, above which the sultan's lengthy official title was inscribed in golden script. This gate led into the first of the four great courtyards of the palace, each of them more sacred than the last. The outer courtyard, through which all visitors to the inner portions of the palace had to pass, was open to all the sultan's subjects and seethed with an indescribable mass of humanity. Any Turk had the right to petition for redress of his grievances, and several hundred agitated citizens usually surrounded the kiosks where harassed scribes took down their complaints. Elsewhere within the same courtyard stood several armouries and magazines, the buildings of the imperial mint and various other arms of the Ottoman government, even stables for three thousand horses. Also present were a pair of white marble pillars on which were placed the severed heads of notables who had somehow offended the sultan, stuffed with cotton if they had once been viziers, or straw if they happened to have been lesser men. Reminders of the sporadic mass executions ordered by the sultan were occasionally piled by the entrance gate as an additional warning: severed noses, ears and tongues.

A sturdy double gate led from this circle of hell into the second, quieter court, forbidden to all but Ottoman functionaries, soldiers and important visitors. This courtyard held the Hall of the Divan – the Turkish council chamber, where the sultan lay on a sumptuous chaise longue, concealed from the gaze of his subjects by a shimmering green silk curtain, to hear the reports of his senior officials or receive the ambassadors of foreign powers. Beyond this second court, and through a third gateway known as the Gate of Felicity, lay the monarch's private chambers and the imperial harem, guarded by black eunuchs brought to Istanbul from Africa. This third courtyard was a place so sacred that no Westerner, and practically no Ottoman, could claim to have set foot in it for almost a hundred years after it was built. Finally, a fourth locked double gateway led from the seraglio into the imperial gardens, which lay at the extreme end of the entire palace complex and commanded magnificent views across the glinting waters of the Bosphorus. Their position at the very heart of the principal symbol of Ottoman power underlined the regard the Turks had for their plants and flowers.

The grounds of the Topkapi were not merely magnificent but extensive. The enormous palace complex contained every sort of garden, as well as flower beds and fountains, pools and orchards. The imposing second court, where the Turks' elite troops assembled each month to be paid in cash from great sacks of money, even contained extensive areas of woodland where deer wandered between the cypress trees and across shaded walks, while to the north of the palace, where the land sloped down to the famous harbour known as the Golden Horn, the gardens extended beyond the walls all the way down to the water.

Flower beds were planted chiefly in the fourth court, where they were often enjoyed by the sultan alone. The only windows that overlooked them were those of the Treasury and a building called the Hall of the Pantry, which housed the royal larders; and these could be shuttered if the Grand Turk so decreed. The gardens of the fourth court were the sultan's principal retreat from the cares of state, and successive monarchs vied with each other to make them ever more beautiful. The rose, the carnation, the hyacinth, the narcissus and, of course, the tulip were all planted in great profusion in this part of the grounds, and particularly on the slopes which led to the highest point of the whole Topkapi complex, a hillock at the northern end which commanded unrivalled views across the Bosphorus and the Sea of Marmara. Upon this promontory, and elsewhere in the gardens, the Ottomans built wooden pavilions called kiosks. They could be used as meeting places or as the focal points of festivals, but were also furnished with solitary divans positioned to catch each passing breeze and offer breathtaking views when the gardens were in flower. Here, more than at any time in his crowded and often violent life, an Ottoman sultan might feel alone and at peace.

Everything about the Abode of Bliss was designed to impress visitors with the extent of Turkish power. The palace's scale was tremendous, its architecture was magisterial, its apartments were decorated in the most opulent fashion possible. Even the most cosmopolitan European merchants would have been awed by the stream of supplies required to feed the imperial court: cartloads of rice, sugar, peas, lentils, pepper, coffee, sena and macaroons all trundled through the Topkapi's gateways, as well as plums preserved in lemon juice, 199,000 hens and 780 wagons of snow each year.

In Suleyman's time, no fewer than five thousand servants toiled among the four courtyards. They ranged from humble watchmen to exotic specialists such as the Chief Turban Folder and the Chief Attendant of the Napkin, whose staff in turn included a full-time Pickle-Server. Among these servants of the sultan were a considerable body of gardeners, the *bostancis*, almost a thousand strong. Their duties in the palace were actually many and varied, and extended far beyond weeding the sultan's tulips – though certainly they performed that function too. *Bostancis* worked as guards, porters and removers of refuse. The five thousand additional members of the corps who worked outside the Topkapi itself formed a royal bodyguard and acted as makeshift police and customs men around the capital.

Most unusually of all, the *bostancis* doubled as the sultan's executioners. It was the royal gardeners who sewed condemned women into weighted sacks and dropped them into the Bosphorus, and the tread of an approaching group of red skull-capped *bostancis*, wearing their traditional uniform of white muslin breeches and cut-off shirts exposing muscular chests and arms, heralded death by ritual strangulation for many thousands of Ottoman subjects down the years.

When very senior officials were sentenced to death, they would be dealt with by the sultan's head gardener, the *bostanci-basha*, in person. The *bostanci-basha* also held the post of chief executioner, and he was required to play a leading role in what was surely one of the most peculiar customs known to history. This was the race held between a con-

demned notable – a deposed vizier or a chief eunuch – and
the man commanded to kill him. As soon as sentence of
death had been passed, it was the practice to allow the
condemned man to run as fast as he was able the half-mile
or so through the gardens and down to the Fish-House
Gate, which stood at the extreme southern end of the
Topkapi and was the appointed place of execution. If he
reached the Fish-House before the head gardener, his sentence was commuted to mere banishment. If, on the other
hand, the condemned man found the *bostanci-basha* waiting
for him at the gate, he was summarily executed and his
body hurled into the sea.*

One of the *bostancis'* less fearsome duties was the provision
of cut flowers to decorate the living quarters of the palace.
In general, the Turks rarely displayed plants in this way,
preferring to leave them in the gardens in which they were
grown. But the custom flourished within the walls of the
Abode of Bliss. Paintings show the sultans' favoured rooms
brightened by a profusion of flowers displayed singly or,
more rarely, in small groups. Tulips, of course, featured
heavily in such arrangements. They were placed in fine glass
vases which were often embellished with filigree using a
technique known as *Cesm-i-Bulbul* – 'the Nightingale's Eye' –
and scattered about a series of low tables.

It was thus, in all likelihood, that Westerners first encountered the cultivated tulips of Istanbul. They came as ambassadors and envoys first, responding to the terrifying successes
that Suleyman's armies enjoyed as they captured Rhodes,
the apparently impregnable stronghold of the crusading

* The last man to save his neck by winning this life-or-death race was the Grand
Vizier Haji Salih Pasha in 1822-3.

Knights of St John, in 1522, then crushed the armies of the King of Hungary in 1526 and besieged Vienna three years later. This glorious string of victories elevated the Ottomans to the rank of the greatest power in the Mediterranean and forced the Christian monarchs of Europe to negotiate with them. Later, mercenaries and merchants also made their way to Istanbul, to enlist with the Turks or seek permission to trade with them. It was one of the minor consequences of the rise of Ottoman power that by the time of Suleyman's death, in 1566, many hundreds of travellers such as these had journeyed to Turkey, a country which had for several centuries been all but closed to the West.

The Westerners found much to remark on. Everything about the Ottoman Empire seemed exotic, from the rowdy vigour of the bazaar to the sensuous grace of Istanbul's mosques. The Turks' passion for flowers, and the remarkable skill with which they tended them, were also among the novelties that drew comment; even the cultivation of plants purely for their beauty seemed strange to visitors from sixteenth-century Europe who were generally accustomed to think of them as things to eat or pound into primitive herbal medicines.

The slender, intensely coloured tulips displayed in every fashionable garden could not fail to attract attention. Whether the travellers who found themselves gazing on the splendid Ottoman gardens were ambassadors or army officers, whether they loved flowers or were indifferent to them, it would have been clear to them that the Turks favoured this one bloom above all others.

By the middle of the sixteenth century, at the latest, the tulip had come at last to Europe's notice. It was ready to resume its journey west.

CHAPTER 3

Stranger from the East

The sailing ships that limped into Goa, the capital of the
Portuguese possessions in India, late in October 1529 were
in a very sorry state. They were badly battered about
and manned by almost literally skeleton crews, having lost
upwards of two thousand men to a combination of fever
and starvation on the long voyage out from Lisbon. The
commander of the flotilla, a noble named Nunho da Cunha,
had survived, however – and his arrival was extremely bad
news for Lopo Vaz de Sampayo, the governor of Portuguese
India.

Da Cunha carried instructions from the King of Portugal
which named him governor in place of Lopo Vaz. Worse,
Vaz himself was summoned home in disgrace. The recall
had been ordered because word had finally reached Lisbon
of how Vaz had usurped the royal favourite who was
supposed to have been appointed governor, and ruled the
Portuguese enclaves on India's west coast for two years in his
stead. Lopo Vaz returned home a prisoner, and languished in

gaol until 1532, when he was banished to Africa to await an eventual pardon.

All this matters because Lopo Vaz de Sampayo is said to be the man who introduced the tulip to western Europe. The horticulturalist Charles de la Chesnée Monstereul, in his *Le Floriste François*, published in 1654, says that Vaz brought the tulip home with him from Ceylon, and several other seventeenth-century authorities make an identical claim.

It is, however, difficult to see how Lopo Vaz could have accomplished this feat. To begin with, tulips do not grow in Ceylon, and the island is hundreds of miles off the route Portuguese ships took when they were sailing home. And though it would not be unreasonable to suppose that the Portuguese in Goa had acquired the flower – either from the Persians they sometimes dealt with in the Gulf, or from Indians who had them from one of Babur's gardens in the north of the subcontinent – the voyage to Lisbon was an arduous one which took about six months when the conditions were good, and anything up to two and a half years when they were not.

If the story about Lopo Vaz is true, then he must have been a tulip maniac of some distinction: keen enough on flowers to persuade his captors to allow him to take his bulbs on board and perhaps even cultivate them in pots on the appallingly crowded and squalid little ships that the Portuguese used to sail to India and back. This is not quite impossible; prisoners of quality got decent treatment in those days whatever their crimes, and Vaz was certainly not carried back to Lisbon in chains. But it is improbable enough for us to doubt that this undistinguished and unlucky noble really deserves to be remembered as the man who first brought the tulip to Europe.

The truth is that no one knows exactly how or where or when the flower first left Asia. The Turks and Persians grew so many of the flowers, and the bulbs were so eminently portable, that it would be very surprising if at least a handful of tulips did not find their way west at some point during the Middle Ages. If they did, however, there is no record of them in the illustrations or chronicles produced at the time, so they can hardly have been planted in quantity, or spread far, and the same applies to any specimens which may have come to Portugal from India; when European botanists did encounter the tulip in the 1560s they thought the flower a great novelty.

Occasionally, some new piece of evidence pointing to the tulip's presence in Europe before the mid-sixteenth century is uncovered, but none has gone without challenge. There is, for example, the problem of the wild red and yellow tulips of the species *T. silvestris* and *T. australis*, which still grow wild in Savoy. It has been suggested that these are survivors of an indigenous European wild tulip which was once linked to its Asiatic cousin by colonies strung across the Balkans. The Savoy tulips, however, have an erratic distribution and are generally found on cultivated land, which suggests that their forebears were planted by people. Then there is a painting of *Virgin with Child*, showing Mary turning her face to flowers which include garden tulips; this was once attributed to Leonardo da Vinci, but it has now been reassigned to his pupil Melzi, who did not die until 1572. Most remarkably of all, there is a Roman mosaic dating to before 430 AD, an exhibit in the Vatican Museum, which unarguably shows a basket of broad-petalled red tulips. Their arrangement is, however, so evidently eighteenth-century in style that it seems the mosaic must have undergone major reconstruction after it was removed from a villa in the suburbs of Rome in the 1700s.

Perhaps the first European to appreciate the beauty of the tulip, then, was Ogier Ghislain de Busbecq, the bastard son of a Flemish lord who was for years the most influential Netherlander at the Austrian court, and who is generally credited with introducing the flower to the West. Busbecq went in November 1554 to Istanbul as the ambassador of the Holy Roman emperor, and remained in the Ottoman Empire for almost eight years, making only occasional journeys home. When he did eventually return, browned by the sun but still bushy-bearded and craggy-eyebrowed after the fashion of the time, he published, in 1581, a book of recollections, written in the form of letters, which described his experiences among the Turks. The letters were packed with intimate and gossipy details, and made his name – both in his own day and among historians who still rely on Busbecq to add colour to accounts of daily life at the height of Ottoman rule. They also contain his own account of how he first encountered the tulip.

Busbecq was travelling overland from Vienna to Istanbul, and had just left the Thracian city of Adrianople for the Turkish capital when he found the flower growing in the wild. 'We set out', the ambassador wrote in one letter,

> on the last stage of our journey to Constantinople, which was now close at hand. As we passed through this district we everywhere came across quantities of flowers – narcissi,

hyacinths, and tulipans, as the Turks call them. We were surprised to find them flowering in midwinter, scarcely a favourable season.

There is an abundance of narcissi and hyacinths in Greece, and they possess so wonderful a scent that a large quantity of them causes a headache in those who are unaccustomed to such an odour. The tulip has little or no scent, but it is admired for its beauty and the variety of its colours. The Turks are very fond of flowers, and though they are otherwise anything but extravagant, they do not hesitate to pay several *aspres** for a fine blossom.

Indeed, Busbecq complained, when he did reach the capital and was presented with some fine tulips by his hosts, 'these flowers, although they were gifts, cost me a great deal, for I had to pay several *aspres* in return for them'. (Another traveller, George Sandys – a son of the archbishop of York – found the Turks equally anxious to press their precious flowers on strangers at about this time, though he was even less enamoured with the gifts than was Busbecq. 'You cannot stirre abroad,' the Englishman grumbled, 'but you shall be presented by the Dervishes and Janizaries with tulips and trifles.')

For many years it was thought that this account of Busbecq's was contemporary and referred to his initial journey to Istanbul, undertaken during the winter of 1554. More recently, however, it has been shown that all the letters which make up his book were written long after the fact – probably not until the early 1580s, when the tulip had

* A Turkish coin.

become reasonably well known in Europe – and that the journey he described could not have been his first, made in the depths of winter. Tulips do not flower at that time of year, even in the Ottoman domain; Busbecq, therefore, must have been misremembering the details of a second journey out to Istanbul which he undertook when the tulips *were* in flower – in March 1558.

Given this revision, it is clear that even if the ambassador's account is accurate in its other details, it would be all but impossible to attribute to Busbecq the tulip's introduction to Europe; the flower was definitely growing in at least one German garden by April 1559. For this to have been Busbecq's work, he would have had to have sent tulip bulbs back within a few months of his arrival for immediate planting that same autumn – which is possible, but not especially likely. It is true that Busbecq did post valuable flower bulbs and seeds from Istanbul to Europe, but as it is not known for certain that he did so any earlier than 1573, it would be dangerous to attribute the existence of one particular tulip to his efforts.

A similar confusion unfortunately swirls about the part Busbecq may have played in giving the flower its familiar name. He is generally believed to have described it as a 'tulipan' because of the petals' resemblance to a folded turban – *dulbend* to the Turks, and *tulband* to the people of the Netherlands. A comparison of this sort could well explain how the word 'tulip' entered the English language, but it cannot have been as a consequence of Busbecq's travels. The term has been traced back as far as 1578, when it appeared in a translation of a botanical work which had originally been published in Latin, so it was certainly in circulation before the ambassador published his famous letters. It took time, in any case, for the word 'tulip' to be generally accepted; in the late sixteenth century European botanists most often referred to the flower as 'lilionarcissus', emphasizing its kinship with more familiar bulbous plants.

It was in the year 1559, then, that the first tulip definitely known to have flowered in Europe appeared. It grew in the garden of a certain Johann Heinrich Herwart, a councillor of Augsburg, in Bavaria. The town was part of the Holy Roman Empire – that remarkable agglomeration of German cities and states which endured from the Dark Ages until its dissolution at the hands of Napoleon, and of which it is important to remember only, in Voltaire's famous phrase, that 'it was neither holy, nor Roman, nor an empire' – and Herwart's garden appears to have been one of its principal adornments. It was certainly well enough known to attract visitors from some distance away.

One of these who came and saw the new flower Herwart had grown was a natural scientist called Conrad Gesner, who lived in Zürich. Like many scholars of that time, Gesner was a polymath who studied zoology as well as botany, and was also a doctor of medicine; one of his most remarkable cases concerned a mysterious epidemic during which snakes and newts were seen crawling from the stomachs of the recently dead. By the late 1550s, he was already compiling the important works of natural history for which he is best remembered, including a comprehensive botany called the *Catalogus plantarum*. He was, in short, well able to understand the significance of the brilliant import he found in Herwart's flower beds.

'In the month of April 1559,' Gesner recalled, 'I saw this plant displayed, sprung from a seed which had come from Byzantia* or as others say from Cappadocia.† It was flowering with a single beautifully red flower, large, like a red lily, formed of eight petals of which four were outside and the rest within. It had a very sweet soft and subtle scent which soon disappeared.' The sketch that Gesner made of his short-stemmed scarlet flower still survives, covered with scribbled marginal notes and queries which pay mute tribute to his interrogative mind. It shows a comfortably rounded flower with tight-wrapped petals, curling delicately outwards at the tips. (There are only six petals, the normal number, in this watercolour, rather than the eight Gesner mentioned in his written description, leaving open the interesting question of whether this pioneer tulip was actually a 'sport', or mutation.) Gesner called it *Tulipa turcarum*, acknowledging that its provenance was the Ottoman realm.

By the time the Swiss scientist completed his sketch in the spring of 1559, however, the tulip had certainly established itself elsewhere in Europe. Gesner himself had already seen a sketch of another specimen, yellow this time, which may have grown in northern Italy; it had been sent to him by his correspondent Johann Kentmann, an artist who had lived in Padua, Venice and Bologna between 1549 and 1551. From these bases, and maybe others, the flower spread quickly from country to country. Its novelty, delicacy and

---

* i.e. Byzantium, that is, the Ottoman Empire.
† A province in central Anatolia, still today the home of flourishing colonies of wild tulips.

beauty made it welcome everywhere, and its wide distribution was assisted by the easy portability of its bulbs.

The time was now right for the tulip. With the discovery of silver mines in the Americas and trade routes to the Indies, there was more money about in Europe than ever before, and the rich were looking for interesting new ways to spend it. The Renaissance had reawakened interest in science, and printing had made both new discoveries and hoarded stores of older knowledge widely available. One consequence of these developments was that botany and gardening were greatly in fashion among the elite. Many of the most influential and affluent citizens of Europe planted their own gardens, and wanted to stock them with rare and coveted plants. Even in Augsburg, Councillor Herwart's collection was easily overshadowed by the gardens of the Fuggers, the fabulously rich Bavarian family of bankers who were to the fifteenth century what the Rothschilds and the Rockefellers were to the twentieth. The Fuggers were growing tulips in Augsburg by the beginning of the 1570s.

There were tulips in Vienna by 1572. They were in Frankfurt by 1593, and reached the south of France by 1598 (possibly much earlier). Bulbs were sent to England too, as early as 1582, where they were soon grown in great quantity. Before the end of the sixteenth century, endless ranks of new hybrids, each more colourful than the last, had already begun to make their appearance: James Garret, one of the best-known botanists in England, spent two decades producing new varieties – so many that even his friend John Gerard (the curator of the physic garden of the London College of Physicians, who mentions them in a *Herbal* published in 1597) confessed that 'to describe them particularlie were to roule Sisiphus's stone, or number the sandes'.

Garret was a Flemish immigrant who worked as a pharmacist and kept a garden at London Wall. His tulips – Gerard mentions that he grew yellow, white, red and lilac varieties – were valued not so much for their beauty as for their supposed medicinal properties. An English botanist, John Parkinson, whose notable treatise on flowers appeared some three decades later, mentions that they could be mashed into red wine and drunk as a cure for 'a cricke in the necke'. They were the stock from which many more varieties were grown; by the reign of Charles I (1625–49), and with the help of imports from the East, more than fifty different tulips were cultivated in the royal gardens.

Gerard might not have been able to catalogue all these flowers, but someone had to. The profusion of new varieties which made the flower unique included tulips which differed one from another not merely in their colour but in their height, the shape of their leaves, and whether they bloomed early or late each spring. What the flower needed now, more than anything, was a man who could create some order from the impending chaos. Without a sound system of classification, the whole genus could get mired in a botanical muddle from which it might never emerge. Without a system of evaluation, moreover – one that indicated which flowers were rare and covetable, and which common and worthless – a trade in tulips could never have developed.

Fortunately, such a man existed. He was indisputably the greatest botanist of the sixteenth century, indeed one of the greatest of all time. He was to become, in important respects, the father of the tulip. His name was Carolus Clusius.

CHAPTER 4

Clusius

One day in the autumn of 1562, a ship sailed into the harbour at Antwerp carrying a cargo of cloth from Istanbul. Somewhere among the bales of Eastern fabrics, consigned to one of the great merchants of the town, were tulip bulbs, perhaps the first ever seen in this part of northern Europe.

The Flemish merchant who had ordered the cloth was surprised to find that his consignment included a package of bulbs. Perhaps they had been intended as a gift, stuffed in among the fabrics by a grateful Ottoman who was making a decent profit on the shipment; at any rate, the merchant had not been expecting them and did not want them. He did not even know what they were. Thinking the bulbs must be some strange Turkish sort of onion, he had most of them roasted and ate them for his supper, seasoned with oil and vinegar. The rest he planted in his vegetable patch, next to the cabbages.

Thus it was that in the spring of 1563, a few strange flowers poked their heads above the dung and detritus of an Antwerp kitchen garden – somewhat to the disgust of the garden's owner, who had been looking forward to another meal or two of Turkish onions. The petals were vibrant red and yellow in colour, and stood out in their delicacy and elegance from the drab leaves of the root vegetables that surrounded them. These fortunate survivors of the cloth dealer's dinner may well have been the first tulips to flower in the Netherlands, and even the Flemish merchant guessed that the latest products of his cabbage patch were something out of the ordinary. He had never seen plants like them before, and, his interest piqued, a day or two later he took a visitor out into his garden and asked him what they were.

The visitor was Joris Rye, a businessman from the nearby town of Mechelen whom the cloth merchant knew to take a keen interest in horticulture. Almost certainly, Rye did not recognize the flowers either; tulips were still all but unknown in northern Europe at this time and Gesner's description of them had yet to be published. Nevertheless, the cloth merchant's visitor was one of the few men in Antwerp who would have understood the importance of preserving the unusual new red and yellow flowers he was shown that day. He was an enthusiastic botanist who filled his own garden back in Mechelen with rare breeds of plants and maintained an extensive correspondence with many of the most prominent horticulturists of the day. So when, with his friend's permission, Rye transplanted the surviving tulip bulbs from the cabbage patch to Mechelen, he did more than just plant them and cultivate them; he wrote to tell his scientific friends what he had found, and asked for their assistance and advice.

One of Joris Rye's most enthusiastic correspondents was

Carolus Clusius, an exceptionally able botanist in his late thirties who had already spent many years travelling through Europe searching for rare and valuable plants. If Rye was going to tell anyone about his new discovery, it would probably be him. So it may well have been in 1563 that Clusius first heard about the tulip.

Clusius was not his real name. He was born Charles de L'Escluse in the French city of Arras in February 1526. His mother was a goldsmith's daughter and his father an extremely minor member of the nobility, whose lordship at Waténes was so poor that he had been forced to take an administrative job at a monastery at St Vaast to help support his family. This proved to be a piece of good fortune, so far as young Charles was concerned, because at a time when many young aristocrats spent more time learning how to hunt and fight than they did in the classroom, it meant he attended the monastery school and received a thorough education.

De L'Escluse proved to be an able student. From St Vaast he went on to the highly regarded Latin School at Ghent, and then to Louvain, which at that time possessed the only university in the Netherlands. He learned Flemish, Greek and Latin and – in accordance with his father's wishes – studied law, receiving his degree in 1548. But at Louvain de L'Escluse did more than learn of legal precedents. It was almost certainly there that he first encountered the Protestant heresy, which Martin Luther and his followers had been spreading across northern Europe. Despite, or perhaps because of, his monastic upbringing, de L'Escluse found Luther's arguments persuasive and he abandoned Catholicism. This meant he no longer felt safe at Louvain, and thus it proved to be the second great turning point in his life.

It is easy, now, to underestimate the significance of de L'Escluse's conversion, and it is important to remember that religion, in the mid-sixteenth century, remained firmly at the centre of both public and private life. To turn one's back on Rome meant risking not only the wrath of the church, which taught that heretics could expect nothing but damnation, but also of Europe's Catholic monarchs, who, with the Inquisition's help, often did what they could to ensure that Protestants entered into eternal life sooner than they expected. Louvain was one of the dominions of the Holy Roman Emperor Charles V, whose possessions stretched from Germany to Spain – a man so pious that he ended his reign by becoming a Catholic monk. That meant de L'Escluse was in very real danger. During one period of persecution, his own uncle was burned at the stake for embracing the same heresy that he now professed. De L'Escluse decided he would be better off in the Protestant lands.

Not daring to tell his staunchly Catholic father where he was going, he journeyed to the town of Marburg, where the local German princeling, Philip the Magnanimous, the Landgrave of Hesse, had recently founded a university specifically to educate the burgeoning Lutheran elite. De L'Escluse enrolled there with the intention of reading law. But while at Marburg he found himself increasingly attracted to the study of botany, and he began to take long walks around the local countryside, searching for rare and unusual plants.

At this time botany was not regarded as a distinct subject worthy of study in its own right. It existed only as a branch of medicine, and then merely as an aid to identifying medicinal plants and herbs. In order to pursue his interest in botany, de L'Escluse had to abandon law and become a student of medicine. This he did, in the summer of 1549. It was also at about this time that he Latinized his name to Carolus Clusius.

De L'Escluse's decision to become Clusius strongly suggests that his espousal of Lutheranism had more to do with an acquired distaste for the Catholic religion than to any great faith in the new ideas. Latin names were in fashion among the humanists – those who rejected old-fashioned and claustrophobic religious authority in favour of the rediscovery of the secular ideals of the classical age. Clusius's passion for botany, and the willingness he showed to move from Catholic lands to Protestant ones and back again in pursuit of his beloved plants, mark him out as a humanist first and foremost.

For the rest of his life, Clusius travelled almost incessantly. He studied in Montpellier, Antwerp and Paris, and spent months tramping through Provence and Spain and Portugal in search of new plants. He went to England, where he met Sir Francis Drake. At the same time he began to earn a reputation as a scientist, publishing books on medicine and pharmacy and entering into what became a prodigious and lifelong correspondence with fellow botanists throughout Europe. It has been estimated that Clusius wrote as many as four thousand letters during his lifetime, an astonishing quantity in an age when the posts were not only slow and unreliable, but also expensive enough to consume a large portion of the botanist's meagre income. He was a natural choice to receive a letter from Joris Rye.

When Rye's first crop of tulips flowered in 1564, Clusius was in Spain on another of his protracted botanical field trips. But twelve months later he was back in the Netherlands, and it may have been in this year that he saw the flower himself for the first time. This is not certain, since he did not mention tulips in any of his writings before 1570, but it can hardly have happened any later than 1568, when the botanist actually moved to Rye's home town, Mechelen, to live with his friend Jean de Brancion. Clusius was quick to recognize the importance of Rye's discovery, acknowledging that the brilliant new flowers 'bring pleasure to our eyes by their charming variety'. But he remained a man of science first and foremost, and when he heard from Rye that the tulips' original owner had eaten them with relish, he resolved to investigate their potential as a foodstuff. He had a Frankfurt apothecary named Müler preserve some bulbs in sugar and then ate them as sweetmeats. They proved, in his considered opinion, to be far tastier than orchids.

Even in a war-torn and frequently starving Europe, tulips never really caught on as a delicacy, perhaps because of their very bitter, onion-like taste (though they were consumed in quantity by the Dutch during the 'hunger winter' they endured at the end of the Second World War). The central role that Clusius played in the history of the flower had nothing to do with the experiments he conducted with Müler and everything to do with his habit of dispatching specimens of the plants he encountered to correspondents all over Europe. Slow as the European mails of the time were, they were unlikely to harm tulip bulbs, and thanks in large part to Clusius and his circle the flower established itself in gardens from Jena to Vienna, Hungary to Hesse.

By now the botanist was at the height of his powers. A contemporary portrait shows a long-faced gentleman of obvious intelligence who has a steady, piercing gaze. He appears handsome and distinguished, and wears his hair brushed back from his forehead, his moustache thick and his beard short and neatly clipped above a full ruff in the fashion of the day. For a solitary man who never married and had minimal contact with his family for years on end, Clusius had a remarkable number of friends. In person he was both earnest and – being often troubled by ill health – inclined to melancholy, yet there was obviously something very compelling about him for he maintained lifelong friendships with dozens of men and women from very different backgrounds. His skill with languages must have been a help – he spoke at least nine, including French, Flemish, Italian, English, Spanish, German and Latin – but it was undoubtedly his passion for plants and extraordinary knowledge of botany that made so many people in so many different countries look forward to his next letter and anticipate the wonders the package might contain. His correspondent Marie de Brimeu, whose proper title was Princess de Chimay and who lived at The Hague, seems to have harboured something approaching maternal feelings for the old bachelor and sent him numerous presents and parcels of food. It was Marie who bestowed on Clusius perhaps the compliment he treasured most: he was, she wrote, 'the father of every beautiful garden in this land'.

Clusius was not the only botanist spreading bulbs and seeds about the continent in this way. For example, some of the tulips he grew for himself in Mechelen had been sent to him by his friend Thomas Rehdiger from Padua. But Clusius was probably the most active, not least because his repeated and lengthy absences abroad meant he rarely kept a garden of his own. Instead, he took pleasure in stocking the gardens of his friends and they in turn provided him with a host of experimental seedbeds with which to investigate the properties of the plants he had discovered.

Clusius made full use of his friends' gardens in preparing some of the masterful botanical studies to which he devoted much of the latter part of his life. These books, which included detailed studies of the flora of Spain, Austria and Provence, were among the first to assume that plants were more than simply potential ingredients in the dubious medical preparations of the day and were worthy of study in their own right. Because of this, Clusius has always been considered one of the founders of botany, not least because he developed a system for classifying plants in groups according to their characteristics – an idea which would later be taken up by Carl Linnæus and turned into one of the foundation stones of modern science.

In May 1573, while Clusius was still living in Mechelen and busy distributing tulip bulbs and other plants throughout Europe, he was asked by the Holy Roman Emperor Maximilian II to go to Vienna and establish an imperial *hortus* – or botanical garden – there. This was a tempting offer. Clusius's father, whom he had been supporting, had just died at the age of eighty-one, freeing his son from the burden of caring for him. The proposed salary of 500 Rhine guilders a year would let Clusius live comfortably at last.

(He had been embarrassingly dependent on the charity of friends for years.) And Maximilian wanted a garden to outshine those his princes and nobles had been cultivating. Clusius, whose poverty and scanty claim to the ranks of nobility had left him with something of an inferiority complex, was flattered by the attention and grateful that the emperor offered formal acknowledgement of his status as a noble. In addition, he already knew a little about his prospective patron, who was one of the few emperors ever to show sympathy towards the Protestant faith; his friend and regular correspondent Johannes Crato von Krafftheim was Maximilian's personal physician. The reports he received were positive and the task certainly seemed an interesting one. So he accepted the proposal.

Today Vienna is a central European city noted for its culture. In Clusius's day, though, it was very much a frontier town. Although one of the most important cities of the Holy Roman Empire and the home of the imperial court, it was also only 50 miles from the Ottoman border and known, not merely to the Empire, as 'the frontline of Christendom'. Under Suleyman, the Turks had laid siege to Vienna with a quarter of a million men in 1529, and they would return again in 1683. So for all the elegance of the imperial residence, the palace of Schönbrunn, the beauty of the broad sweep of the Danube, and the bustle of the narrow, crowded streets in the centre of the town, the state of the gates and the walls mattered more than the addition of a few flower beds. Gardens were something of a luxury.

From the moment Clusius arrived, he discovered that while there were advantages to working for an emperor, his job was attended by many frustrations. Maximilian was busy and Clusius had to wait two months for an audience and more than a year for any sign of activity at the site chosen for the garden. Worse, the imperial chamberlain in charge of the finances for the *hortus* and Clusius's own pay turned out to be a strict Catholic who made life as difficult as he could for the Protestant botanist. On the other hand, Clusius did begin to receive regular parcels containing the bulbs and seeds of many plants from the imperial ambassador at Istanbul, and struck up a botanical friendship with Ogier Ghislain de Busbecq, who was now back at court. The two men exchanged presents of plants, and when Busbecq left for France in 1573, he presented his friend with a large quantity of seed.

Clusius did not get the opportunity to plant it for another two or three years, by which time Busbecq's gift had shrivelled so badly he feared the seed was dead; but it did germinate eventually and turned into a spectacular profusion of tulips – a suitable mark indeed of the friendship between two champions of the flower.

For all that, the garden project continued to languish and by the summer of 1576 Clusius's pay was eleven months in arrears. Then Maximilian died suddenly, and matters took a turn for the worse. The new emperor, Rudolf II, was a Catholic zealot who dismissed every Protestant serving at his court. Worse, he had little interest in flowers, and the fledgling *hortus* was torn up so the land could be turned into a riding school. Clusius was horrified. Although his services were always in demand, he never in his life worked for another monarch.

He stayed on in Vienna for a while, disillusioned and plagued by repeated thefts of rare plants from the private garden he still maintained there. At a time when much-coveted plants might exist in only one or two gardens in the whole of Europe, the organized theft of specimens was, if not exactly commonplace, then certainly far from unknown. Like antique thieves today, the men who carried out these crimes were often knowledgeable connoisseurs and knew exactly what they were looking for. (Those who didn't simply bribed the ill-paid servants who acted as gardeners for the necessary information.) The plant thieves generally worked for nobles and merchants who wanted an enviable garden of their own for the minimum of effort. Miscreants such as these seldom made any effort to conceal the evidence of their activities, but there were no police to investigate such crimes and the authorities were not remotely interested in prosecuting the well-connected for such trivial offences. On at least one occasion, Clusius was forced to grind his teeth while a Viennese noblewoman proudly conducted him around flower beds she had stocked with plants stolen from his garden.

He was a old man now, over sixty and half crippled by a bad fall in his bath. He suffered from undiagnosed stomach complaints and had lost all his teeth; and now that his imperial salary had been stopped, he was badly off again and needed some way to supplement the meagre income from his lordship and the occasional food parcels he received from his friends. He yearned, too, for some sort of academic recognition of his life's work. And, in the end, it came.

CHAPTER 5

Leiden

In January 1592, a large sealed package arrived at the lodging-house where Clusius was living. It was a letter from Marie de Brimeu containing the news that he had been offered a post in the medical faculty of the University of Leiden.

Leiden was a large manufacturing town in the United Provinces of the Netherlands – not a place that Clusius would normally have chosen to live. But de Brimeu's letter arrived at a particularly opportune moment. After leaving Vienna, the old botanist had retreated to Frankfurt to be close to his friend and patron, the Landgrave of Hesse. But the Landgrave had just died, and his heir had cancelled the small yearly pension on which Clusius relied. Deprived of his principal source of income, he badly needed to find work. The post at Leiden offered not just recognition of his life's work but a salary of 750 guilders a year plus his travel expenses; in addition, several of his correspondents already worked at the university, and the man who had actually proposed him for a professorship, Johan van Hoghelande, was a friend with whom he had exchanged flower bulbs for years. After some consideration, and not without reluctance, Clusius decided to accept van Hogheland's offer.

Thus it was that the man who had done more than anyone to popularize the tulip made his way to the Dutch Republic, where the flower would become truly famous. Clusius reached Leiden on 19 October 1593, bringing with him many of his precious plants. Among his baggage was his extensive and by now quite valuable collection of tulip bulbs.

The botanist's new home was a substantial town of perhaps 20,000 people which stood more or less in the centre of the United Provinces. The city was built around the ruins of a medieval castle and was noted as a busy centre of the textile trade. Yet when Clusius arrived there, civic confidence was in a fragile state. Leiden might have been large by Dutch standards, and the university was its pride and joy, but the town was only just emerging from a century of stagnation to embark on a period of rapid expansion which would culminate in its becoming one of the two biggest cloth towns in Christendom. To a casual observer there seemed to be no reason why anyone who lived outside Holland should have known or cared much about Leiden. Yet, as Clusius himself would have been well aware, in the closing years of the sixteenth century Leiden was actually one of the most celebrated places in Europe.

The town's fame rested on the heroic role it had played in one of the defining events of the century: the Dutch revolt. For much of the sixteenth century, all seventeen provinces that made up the Low Countries – both the south, which is now Belgium and Luxembourg, and the north, which became the United Provinces and is now the

Netherlands – were among the ancestral lands of the king of Spain. The king (and between 1556 and 1598 it was the same Philip II who loosed the Spanish Armada on England) was the most powerful monarch in Europe, controlling a vast empire which already included much of South and Central America. He was fighting the Turks in the Mediterranean and the English in the Caribbean, as well as confronting the French in Europe. The southern provinces of the Netherlands were centres of commerce and important strategically in any conflict with France, but the lands to the north were a long way down Spain's list of priorities. Certainly the king was disinclined to listen to protests from the Netherlands about the high rates of tax being imposed to pay for his wars or the presence of large numbers of Spanish troops who were being fed and watered at Dutch expense. As a fervent Catholic, he was even less willing to tolerate the rise of Protestantism in his possessions, and from the 1550s there was considerable persecution of the new religion throughout the seventeen provinces.

By the 1570s, popular feeling was running against Spain in many parts of the Netherlands, but particularly in the seven provinces which lay to the north of the rivers Waal and Maas. These provinces – they were Holland, Zeeland, Gelderland, Utrecht, Groningen, Overijssel and Friesland – were poorer than their ten brothers in the south, but they occupied lands that were difficult to attack. When open revolt finally broke out in 1572, even the vaunted armies of Spain proved incapable of conquering them.

The spark for the revolt was provided, inadvertently, by Queen Elizabeth of England. For several years she had been harbouring a group of Protestant Dutch pirates known as the Sea Beggars in her Channel ports. Under pressure from Spain, she finally expelled them in April 1572 and, with nowhere else to go, the Beggars went marauding along the Netherlands coast until they came upon the little port of Brill. Discovering that it had been temporarily left without its Spanish garrison, they occupied the town to the general acclaim of its inhabitants. Five days later, the Beggars sailed down the Zeeland coast and seized Flushing, a strategically vital port which among other things controlled Antwerp's access to the sea.

From there, revolt spread quickly across the Netherlands. By July almost the whole of the province of Holland, with the exception of Amsterdam, was in the hands of the rebels. At Leiden popular opinion was so in favour of the Beggars that the town went over to the revolt spontaneously, before any Protestant troops could be sent to form a garrison. The citizens of the town chased out the few loyalists there and then thoroughly pillaged the Catholic churches, thus earning the undying enmity of the Spaniards.

One of those who reacted most quickly to the news of the uprising was William the Silent, the Calvinist Prince of Orange, who soon became the main figurehead of the revolt. He had himself proclaimed *stadholder* – a title roughly equivalent to governor – of Holland and then 'Protector' of the Netherlands as a whole. Before long William had put himself at the head of a substantial army and prepared to resist the inevitable Spanish counter-stroke.

It came before the end of the year, and when it did the Spaniards showed that their strategy was to terrorize the Dutch into submission. Several small towns were overrun and their citizens massacred, sometimes almost to the last inhabitant. Fear of the Spanish terror cowed many of the cities that had declared for the prince, and before long only the provinces of Holland and Zeeland remained committed to the revolt. A huge Spanish army gathered to push north into the last rebel territories and snuff out the rebellion. Standing in its way was Leiden.

The siege of Leiden was the hardest-fought, the costliest and the most decisive of all the actions of the revolt. Had the town fallen, the Spaniards probably would have succeeded in mopping up the remaining Dutch resistance and restoring their rule throughout the northern provinces. The Dutch Republic would have been still-born, trade and commerce would have remained concentrated in the south, the wealth generated by overseas trade would never have flooded into Holland and the tulip mania could not have taken place.

As it was, Leiden prevailed, but only after a desperate siege which lasted four months. At the end of that time the citizens had run out of food and, in a last bid to save the town, the *stadholder* ordered the dykes along the Maas to be cut so the river waters would flood the land around the town and drive out the besiegers. The waters did rise, but not so far as to end the siege. Then, in what the pious Dutchmen considered a direct intervention by the Almighty, the wind changed direction, a huge storm arose, heavy rain fell and the river waters surged forward until the Spanish soldiers were forced to flee and the men of the Beggar fleet were able to relieve the town by sailing their ships over what had been farmland only days before.

The epic resistance of Leiden saved the Dutch revolt, but the Spanish threat remained a very real one for decades after the first stage of the uprising was successfully concluded and the seven rebel territories formed themselves into a republic – the United Provinces of the Netherlands – with the Prince of Orange still in the important role of *stadholder* and commander-in-chief. There were several further Spanish invasions of Dutch territory, the last in 1628, and so although the almost incessant conflict was broken by a long truce which ran from 1609 until 1621, the Dutch faced the expense of maintaining armies in the field and the constant threat of another attack until about 1630. From then until Spain was eventually forced to recognize the United Provinces by the 1648 Treaty of Münster, the threat was all but ended and the cost of maintaining a substantial army and navy could be cut. The money that was saved was diverted into the Dutch economy, which flourished as never before after 1630.

When Clusius arrived in Leiden two decades after the drama of the siege, the university there was the only one in the United Provinces. It was also still very new, having been founded only in the spring of 1575. The establishment of such a centre of learning was a necessary step for the new nation to take; not only was it expressly intended as a cultural declaration of independence from Spain, it was also needed to produce ministers for the church and young men fit to govern the United Provinces. At this time most other colleges in Europe gave priority to religious training and in fact the majority of universities were directly controlled by the church, which limited the breadth of the education available. The Dutch government was determined that this should not be the case at Leiden. Teaching was offered in law, medicine, mathematics, history and other humanist subjects as well as theology, and control of the university was vested in seven curators who were nominated not by the church but by the provincial parliament and the burgomasters of Leiden.

All this was much to Clusius's liking, but the young university's humanist policy had caused unexpected problems. Between 1575 and the early 1590s, Leiden's dangerously liberal reputation meant that the leaders of the reformed church looked with suspicion on the graduates of its theology school, and Dutch students who planned to pursue a career in the clergy generally chose to enrol at one of the more strictly Protestant north German universities. The ever-present danger that the United Provinces would fall to a renewed Spanish attack deterred scholars from matriculating in other subjects as well, and in its first dozen years, Leiden recruited no more than 130 theology students all told, and fewer humanists. It took some dramatic Dutch victories and the easing of the military situation in the early 1590s to make the place more attractive to prospective students. The university that Clusius agreed to join, then, although nominally two decades old, was really only just being born when the old botanist finally arrived in the Dutch Republic.

It was a good time to come to Leiden. Suddenly money was available to improve the facilities, hire more staff, buy more books and offer grants to more young scholars. Over the next half-century, the number of students in residence rose fivefold, from a hundred to five hundred, and the library built up one of the most comprehensive collections available anywhere. The university became particularly famous for its School of Anatomy, where human cadavers were dissected. The mysteries of the body were just beginning to be explored in this period, and anatomy was one of the most fashionable subjects of the day. At Leiden, public interest was so great that dissections were frequently carried out before spectators, and visitors were also encouraged to visit the university's anatomical museum, where over the years such wonders as an Egyptian mummy, stuffed tigers, a giant crocodile and an immense whale's penis were put on display. In the fifty years after Clusius's arrival, this sort of excellence resulted in Leiden becoming possibly the best – and certainly the most popular – university in Europe. More students were enrolled there than at either Cambridge or Leipzig, the next two largest establishments in the Protestant north, and Leiden's student body was also more cosmopolitan and international than any of its rivals'.

Clusius benefited as much as anyone from this sudden influx of confidence and funds. His principal task was to establish a *hortus academicus* at Leiden in imitation of the one set up at the University of Pisa in 1543, which had been the first botanical garden in Europe. Since then similar gardens had been established at the universities of Padua, Bologna, Florence and Leipzig, but there were still none in the United Provinces. Leiden's *hortus* was thus an important symbol not just for the university but for all the Dutch Republic, and the garden was amply funded and laid out on a generous scale. When it was complete, it covered nearly a third of an acre and was divided into four main sections, each of which contained some 350 individual beds.

With the memory of his frustrating years in Vienna still fresh, Clusius was particularly pleased with how quickly his *hortus* was laid out and planted. He was by now too infirm to do any of the physical labour involved himself, but the university provided him with a very able assistant in the shape of an apothecary from Delft named Dirck Cluyt, and under Cluyt's direction work on the garden was complete by September 1594, less than a year after Clusius's first arrival at Leiden. It made a pleasant contrast to the dilatoriness of Maximilian and the imperial court.

The speed with which the *hortus* took shape helped to distract Clusius from some of the difficulties of living in Holland. He had to endure the hard winter of 1593–4, during which Leiden's mice made short work of 150 of the precious bulbs in his personal collection, and then the miserable weather that the Low Countries experienced in 1594. It was a year of seemingly constant wind and rain which damaged many of the plants in the botanical garden and did nothing to improve the health of a man who was now sixty-eight years old.

Although he was contractually obliged to look after the garden and to visit it each afternoon in summer to answer the questions of students and distinguished visitors, the elderly Clusius stubbornly refused his new employer's request that he deliver lectures on botany as well. Instead, he devoted much of his time to beekeeping and to pottering about the private garden he had insisted that the curators provide for him. While the *hortus* was largely given over to herbs, medicinal plants and exotic novelties such as the potato – only recently introduced from the New World and still regarded as quite possibly poisonous – Clusius sowed the collection of tulip bulbs he had brought with him from Frankfurt in his own garden, where he continued to cultivate the flower and delve into its mysteries until his death in 1609, at the very advanced age of eighty-three.

Carolus Clusius was without question the most important botanist of his day. He was a true scientist whose greatest works, such as his surveys of the plants of Austria and Spain, remained the standard texts on their subject for more than a century, as well as a pioneer in the most literal sense; the brief history of fungi which he published in 1601 was more or less the first thing that had ever been written on the subject. For the last quarter-century of his life he served as a sort of living *vade mecum* for the botanists and flower-lovers of Europe, keeping up a vast correspondence. This, and the particular interest he had in bulbous plants, ensured that the tulip spread far more rapidly through Europe than might otherwise have been the case. From this point of view he really was, in the words of another valued compliment – this one from the pen of Prince Emanuel of Portugal – 'true monarch of the flowers'.

Yet Clusius's importance, during his final years at Leiden, lay not so much in the bulbs he brought to the university as in how he studied them once they were planted. The old botanist was not the first person to grow tulips in the United Provinces; according to one reliable chronicler, that honour belonged to an Amsterdam apothecary called Walich Ziwertsz.,[*] a Protestant fanatic who is remembered chiefly

* Surnames were still relatively uncommon in the United Provinces in the sixteenth and early seventeenth centuries. Most people still identified themselves using patronymics – Walich Ziwertsz. would have been the son of one Ziwert or Sievert. Because it was unwieldy to spell out the full patronymic, which in this case would be Ziwertszoon ('Ziwert's son'), it was also common practice to abbreviate written names by placing a full stop after the 'z' of 'son'. When spoken, the name would have been pronounced in full.

for his denunciation of the popular custom of celebrating the festival of St Nicholas on 25 December. Ziwertsz. is known to have cultivated tulips in his garden before 1573, when Clusius was still in Vienna. Nor was the master of the *hortus* the first person to raise the flower in Leiden; his own friend Johan van Hoghelande had planted bulbs at the university before his arrival, having received a small stock from Joris Rye. Clusius was, however, the only man in the United Provinces – perhaps in all Europe – who was perfectly qualified to describe and catalogue and understand the flower.

Clusius's first discussion of tulips appeared in his description of Spanish plant life, the *Historia* of 1576. Over the years he amended and expanded this early work, publishing enlarged treatises on the flower in 1583 and finally in his masterpiece, the *Rariorum Plantarum Historia*, which appeared in 1601 while he was still at Leiden. It is largely thanks to these works that we know as much as we do about the early history of the tulip in Europe. Clusius's treatises also included a detailed description of the flowers he had personally encountered or heard of from his many correspondents. Like all contemporary botanists who took an interest in the genus, he was principally impressed by the ease with which new varieties of tulip could be produced. No other flower, he observed – except perhaps the poppy – was remotely as diverse.

Thanks largely to the efforts of the gardeners of Istanbul, the number of tulip variants known in Europe was already substantial in Clusius's day. The botanist was able to catalogue no fewer than thirty-four separate groups, which he classified according to their colour scheme and the shape and arrangement of their leaves and petals. He was also the first to distinguish between early-, mid- and late-flowering tulips, of which the first appear in March and the last not until May.

Working from the solid foundation Clusius provided, later botanists have added considerably to our understanding of the tulip. The flower has now been grouped with other bulbous plants such as the iris, the crocus and the hyacinth, and classified among the Liliaceae. In all, about 120 different species of tulip – and countless individual varieties – have been catalogued to date.

In scientific works, an important dividing line is drawn between what are known as 'botanical tulips', which originate in the wild, and cultivars, which are hybrids reared in the garden. In Clusius's time, the tulips which were produced in the United Provinces were a mixture of wild flowers and an ever-increasing proportion of cultivars, the earliest of which were produced by chance crossings of two botanical tulips. Botanists have been able to identify fourteen different species of wild flowers as the building blocks that produced the flood of Dutch cultivars which adorned the seventeenth century. Not every species played an equal part in creating this diversity. Some botanical tulips produce hybrids more readily than others, and the most malleable species which had found their way to the Dutch Republic included the Persian tulip – today known in Clusius's honour as *T. clusiana* – the tapered tulip, *T. schrenkii*, and the fire tulip, *T. praecox*. Genes from these species were present in a large proportion of the cultivars that excited admiration in the Netherlands, but in truth Dutch tulips had been produced by the crossing of flowers which had come to the United

Provinces from all points east, from Crete to Kurdistan. That was the secret of the tremendous variety they exhibited.

Whether they were botanical species or cultivars, tulips could be grown from either seed or bulbs. Growing from seed is a chancy business. Plants grown from a single pinch of seed gathered from just one flower can exhibit considerable variation, so it is impossible to know exactly what sort of tulip will emerge. Important details such as the colour and the pattern of the flower can only be guessed at, which makes the process frustrating for anyone seeking consistency. And it takes six or seven years to produce a flowering bulb from seed, a very time-consuming process which must have seemed even more so in an age when the average life expectancy at birth was not much more than forty years.

Once a tulip grown from seed has matured and flowered, however, it can also reproduce itself by producing outgrowths, known as offsets, from its bulb. These miniature bulbs are effectively clones of the mother tulip and will produce flowers that are identical to it. Offsets can be separated from the mother bulb by hand, and, in another year or two, become bulbs capable of flowering themselves. From the point of view of both the commercial grower, who seeks consistency, and the gardener, who prefers not to wait seven years to see a flower, propagation through offsets is greatly preferable to raising tulips from seed. However, reliance on outgrowths does have one significant disadvantage: most tulip bulbs will produce only two or three offsets a year, and can do so for only a couple of years before the mother bulb becomes exhausted and dies.

For this reason new varieties of tulip multiply only very

slowly at first. Once a grower has identified, in a single flower of some new variety, great beauty or strength which he may be able to sell, he will – even if all goes well – quite possibly only have two bulbs the next year, four the year after that, eight in the next year and sixteen in the fourth year of cultivation. If he parts with some of these bulbs, moreover, he limits his own ability to produce large quantities of the new variety. Plainly, then, it can take a decade for a new tulip to become available in any sort of numbers – and in Golden Age Holland, where propagation was a poorly understood mystery at best, the numbers of bulbs that were actually produced would have fallen well short of the theoretical maximum. Any rare and coveted variety would thus inevitably remain in short supply for a number of years, and there was nothing that even the most brilliant bulb growers could do to increase production to meet demand.

As soon as tulips of different species are placed close together in gardens, where insects can take pollen from one flower to another, the chance of producing hybrids is substantially increased. And as the new varieties thus created are themselves crossed with other flowers, increasingly elaborate cultivars emerge, bearing the different characteristics of their many forebears. Because tulips of different species do not often grow together naturally, complex hybrids of this sort do not easily occur in the wild. They are, in the strict sense of the word, freaks. But they are also less straightforward, more subtle than wild flowers and thus much sought after by connoisseurs.

The most favoured tulips were those that exhibited the most evenly sculpted petals and the most eye-catching markings. Indeed, Dutch cultivars of the Golden Age were

celebrated and valued far beyond the borders of the Republic
for the elaborate and often riotous colours they exhibited.
By the middle 1630s, no fewer than thirteen groups of
flowers had been created, each with its own distinctive
colour scheme. These ranged from the *Couleren*, which were
simple single-coloured tulips in red or yellow or white, to the
rare *Marquetrinen* – late-flowering varieties which exhibited at
least four colours. The *Couleren* would have been botanical
tulips, or at least cultivars closely related to them, while
Marquetrinen tulips must have been fairly complex hybrids.
The latter were grown mostly in Flanders and France and
do not figure in the records of the tulip mania.

In the Dutch Republic, the most popular of the thirteen
groups were the *Rosen*, the *Violetten* and the *Bizarden*. *Rosen*
varieties, which were by far the most numerous, were
coloured red or pink on a white ground. During the first
third of the seventeenth century, almost four hundred *Rosen*
tulips were created and named. They varied considerably
one from another; some were marked with flames of colour
that were so broad and vivid they all but drowned out the
whiteness of the petals, others bore only the slightest hint
of ruby pigment. Tulip connoisseurs knew many of the
most coveted varieties intimately, and always favoured
flowers which displayed only the merest trace of colour to
those that were swamped by red. The same criteria were
applied to other classes, too. The seventy or so *Violetten*, as
their name suggests, were purple or lilac on white, and the
Bizarden, which on the whole were the least favoured of the
three and existed in only two dozen varieties, were coloured
red, purple or brown on yellow. Varieties that reversed the
usual colour schemes also existed, and were generally classed

with them: for example, *Lacken* tulips were purple flowers with a broad white border and were grouped with the *Violetten*, while the handful of *Ducken* cultivars, which were red with a yellow border, could be found among the *Bizarden*.

It was the patterns that these contrasting colours formed that really excited gardeners, and it is impossible to comprehend what happened later without understanding just how different tulip cultivars were from every other flower known to horticulturalists in the seventeenth century. The colours they exhibited were more intense, more concentrated than those of ordinary plants; mere red became bright scarlet, dull purple a bewitching shade of almost-black. They were also brilliantly defined, quite unlike the indefinite flushing displayed by other multicoloured flowers as their petals shaded gradually from one colour to another.

The distinguishing colours of Dutch tulip cultivars – the reds of *Rosen* tulips or the purples of the *Violetten* – generally appeared as feathers or flames which ran up the centre of each petal and sometimes also formed a border around its edges. These colours occasionally also appeared as mottled patches on the plant's stem, though they never tainted the purity of the flower's base, which was always either white (sometimes tinged with blue) or yellow, depending upon the variety. The patterns were unique to each flower, and though two plants of the same variety might closely resemble each other, they were never absolutely identical.

From the earliest days of the bulb craze, Dutch tulipophiles used the subtle variations of these flames and flares of colour to grade their flowers according to a strict set of criteria. The most highly prized tulips, termed 'superbly fine', were almost entirely white or yellow in colour, displaying their flames of violet, red or brown only in thin stripes which ran along the centre and the edges of their petals. Flowers which in the opinion of the connoisseurs flaunted their bright colours too wantonly were termed 'rude' and were much less cherished.

Botanical tulips are noted for their robust and simple colour schemes, so how did the celebrated cultivars of the Dutch Golden Age become so elaborately coloured? The solution to this problem is simple but disturbing: they were diseased. The great irony of the tulip mania is that the most popular varieties, the ones which changed hands for hundreds or even thousands of guilders, were actually infected with a virus, one apparently unique to tulips. It was this that caused both the spectacular intensity and the variations in the colours of their petals and explained why tulips, alone among the flowers of the garden, displayed the distinct and brilliant colours that collectors came to crave.

Even in Clusius's day, it was obvious that something strange was happening to the tulips grown in Leiden and elsewhere. A bulb which one year had produced a unic-oloured tulip might become a *Rosen* or a *Bizarden* the next. This process was known as 'breaking', and the bulb of a flower which had undergone the process was said to be 'broken' while those which remained unicoloured were called 'breeders'. The whole process was extremely unpredictable. There was no way of telling if or when a flower would break; one tulip might bloom in the spring with a dazzling new array of colours, while another, of the same variety and planted next to the first in the same flower bed, remained quite unaffected. Breaking was common in some years, less so in others. Similarly, a broken bulb might – albeit rarely –

produce an offset which turned out to be a breeder, and no grower could be sure that a breeder bulb would not break. The only certainties seemed to be that tulips grown from seed were invariably breeders, and that, once broken, a mother bulb would never again produce a unicoloured flower.

There were clues here to the nature of the disease, and Clusius was a careful enough observer to notice that broken tulips were slightly smaller and definitely weaker than the flowers produced by breeder bulbs. But at a time when the mechanisms by which diseases are communicated remained unguessed at, the phenomenon of breaking seemed akin to magic to most of his contemporaries. Try as they might, growers could not force a breeder bulb to break when they wanted it to. Some turned to alchemical potions made of pigeon dung, which they applied to the bulbs; others tried cutting the bulbs of two different-coloured tulips in two and binding the opposing halves together in the hope of producing a flower sporting both colours. These devices rarely had the desired effect.

Exactly when the tulip became infected with a virus is not certain. The earliest observations of the phenomenon date to about 1580, but the disease was probably older than that. In truth the plant became vulnerable to disease as soon as it entered a garden; any flowers raised in artificial proximity by humans face threats they do not encounter in the wild. Cultivars may be poorly cared for or discarded in favour of some new favourite, but, in particular, they can pick up diseases to which the more robust botanical species have developed an immunity, or which at least spread more slowly in the wild.

The mystery of breaking remained unsolved until well into the twentieth century, when the agent that causes the disease, which is sometimes called the mosaic virus, was finally identified by staff at the John Innes Horticultural Institute in London. By permitting aphids to feed on broken bulbs and then on breeders, they were able to show that the breeder bulbs visited by the aphids broke twice as often as a control sample – thus simultaneously proving that the disease was caused by a virus and demonstrating the mechanism whereby it was transmitted from one tulip to another. Further experimentation showed that the mosaic virus could infect both a flower when it was growing in a garden, and a bulb which was being stored prior to planting. Perhaps ironically, given the efforts of old Dutch growers to induce breaking by binding half-bulbs together, the method used at the John Innes Institute to persuade aphids to feed alternately on infected and uninfected tulips was to graft halves of broken bulbs on to breeders.

Well before Clusius's death, the broken tulips he grew in his private garden at Leiden were attracting the attention of connoisseurs eager to secure specimens of these unique new flowers for their own gardens. The old botanist soon found himself almost overwhelmed with requests for tulip bulbs. Many, he knew, came from people who merely wanted to follow a burgeoning fashion for the flower, and had no real interest in botany and no idea how to cultivate bulbs; others were from people he suspected of planning to sell his bulbs on for whatever they could get. In any case, his own supplies were not remotely adequate to meet the demand. 'So many ask for them,' he wrote to his friend Justus Lipsius, a humanist scholar who had been one of the pillars of Leiden University in its formative years, 'that if I were to satisfy every demand, I would be completely robbed of my treasures, and others would be rich.'

Unfortunately for Clusius, at least some of those who implored him for bulbs would not take no for an answer. Just as he had been in Vienna, he began to be plagued by repeated thefts from his garden. Twice during the summer of 1596, and again in the spring of 1598, thieves stole tulip bulbs from him while he was away. The total loss must have been substantial, because we know from Clusius's surviving letters that more than a hundred bulbs were taken in just one of these raids, and the old man was so distressed by the loss – and by the familiar lack of interest that the authorities at Leiden showed in investigating the thefts – that he vowed to give up gardening altogether and disperse the rest of his collection among his friends.

Over the years, Clusius's reputation has suffered from the suggestion made by one contemporary chronicler that these thefts occurred because he asked an exorbitant price for his flowers and stubbornly refused to hand over bulbs to anyone who would not meet it. Nothing could be further from the truth. Throughout his long career, the botanist showed great generosity in sending samples of his finds to friends for nothing – 'con amore', as he sometimes put it in his letters – and the only people he refused to supply were those he suspected would not value his gifts. The people who organized the theft of his bulbs at Leiden fell into the latter category, and Clusius was surely right to suspect their motives from the start.

Nevertheless, the thefts did have one positive result.

Clusius's tulips were far from the only ones in the United Provinces in the 1590s, but his collection was certainly the most varied and the best. As a result of the thefts, these precious bulbs were distributed throughout the Netherlands, north and south, and they flourished. In some of their new homes, at least, they must have become the parents of new hybrids, varieties which in their turn bred and formed an important part of the stock of bulbs traded in the next century. The Leiden bulbs thus became the progenitors of the flowers traded later on and thanks in part to them, in the words of the chronicler, 'the seventeen provinces were amply stocked'.

CHAPTER 6

An Adornment to
the Cleavage

From its first discovery, the beautiful colours and endless variations of the tulip marked it as an exceptional flower. There was general agreement on this point, not only between Turks and Dutchmen, but also among botanists of every nationality, and by 1600 it had been widely acclaimed throughout Europe. The tulip was, the French horticulturalist Monstereul wrote a little later, supreme among flowers in the same way that humans were lords of the animals, diamonds eclipsed all other precious stones, and the sun ruled the stars. That judgement, to a seventeenth-century mind, said something important about the tulip. If humans were God's chosen creatures, then the tulip was surely God's chosen flower.

The popularity of the new flower was such that garden-lovers soon began to strive to outdo each other by producing ever more dazzling and brilliantly coloured varieties. Thanks in part to the work of Clusius and his circle of correspondents, a good number of different hybrids were now available; to the tulips of the Netherlands and the dozens of varieties produced by James Garret in England must be added the 41 French cultivars catalogued by the botanist Lobelius and uncounted others elsewhere; certainly many more than 100 in 1600, and 1000 (of which at least 500 were Dutch) by the 1630s. The latter total compares remarkably favourably to the 2500 or so species produced by the mid-eighteenth century and the 5000 cultivars recognized today.

Nevertheless, the number of bulbs available at the turn of the century remained somewhat limited. Most of the new varieties had so far produced only a handful of tulips, and, largely for this reason, the flower remained the passion of the privileged few. It was grown principally by rich connoisseurs, who valued it for its beauty and the intensity of its colours. These men traded prized flowers among themselves, but because they were, almost without exception, wealthy in their own right, they only rarely cared to make substantial profits from these exchanges.

By the end of the sixteenth century, small groups of tulip connoisseurs existed throughout Europe. They could be found in the city-states of northern Italy, in England and in the Holy Roman Empire. But, thanks largely to the early introduction of the tulip to the southern Netherlands, the largest concentration of enthusiasts were to be found in the Low Countries among the members of the Flemish nobility and gentry. Many of these connoisseurs had obtained their first bulbs from Carolus Clusius and his companions. Clusius's colleague Mathias Lobelius published a list of them in 1581; they included Marie de Brimeu, who had a fine garden at her home in The Hague, Joris Rye of Mechelen, and Clusius's lifelong friend Jean de Brancion.

From the Netherlands, the tulip soon spread south to

France, where the soil of Picardy was well suited to the cultivation of bulbs. Around 1610 there was a craze for flowers in Paris and fashionable nobles began competing with each other to present the ladies of the French court with the rarest and most dazzling specimens they could find. When the idea first caught on, most of the blooms exchanged in this way were roses, which had been, for several centuries, by far the most popular garden flowers. But the nobles of the French court found in the tulip something capable of surpassing the reigning empress of the garden. The subtle elegance of the flower – not to mention its novelty and rarity – quickly established it as the new favourite of the court. The fashion for tulips seems to have raged at least until the wedding of the young king Louis XIII in 1615, when aristocratic ladies wore cut flowers as an adornment to the cleavage, pinned to the plunging necklines of their low-cut dresses, and the most beautiful varieties are said to have been as highly esteemed as diamonds. The Dutch horticulturalist Abraham Munting, writing later in the century, recorded that at the height of the French craze a single tulip of especial beauty – and a cut flower, not a bulb – changed hands for the equivalent of 1000 Dutch guilders.

Of course, the nobles of the court soon sought new diversions. But their enthusiasm for the tulip had important consequences, for Parisian society, even in the seventeenth century, was renowned throughout Europe for its elegance and style, and the fashions of the court were taken up and followed elsewhere. Indeed they often continued to flourish in the backwaters of the continent long after the French themselves had moved on to some other craze, and it was not at all uncommon for visitors to the west of Ireland or the forests of Lithuania to find the ladies there dressed in styles that Paris had discarded ten or twenty years before. The passion for tulips which swept through the court of Louis XIII for a few short years thus did much to ensure that the flower would be looked on with high favour throughout the continent for decades to come.

The first people to follow the fashion of the French court were the French themselves. Shortly after the tulip became popular in Paris, a miniature mania for the flower took place in northern France. There are, unfortunately, no contemporary sources of information about this episode, which by all accounts foreshadowed what was later to occur in the United Provinces. If later reports are to be believed, however, the passion for tulips was such that in about 1608, a miller exchanged his mill for a single specimen of a variety called Mère Brune and another enthusiast handed over a brewery valued at 30,000 francs in return for one bulb of the hybrid Tulipe Brasserie. A third account of the same episode tells of a bride whose dowry consisted of one solitary bulb of a new *Rosen* tulip which her father had bred and named, with due sense of occasion, Marriage de Ma Fille. (The groom of this tale is supposed to have been delighted by the magnificence of the gift.) These stories may be apocryphal. It is, however, certain that the fashion for tulips soon spread through the rest of Europe. By 1620 the flower was nowhere more popular than in the United Provinces, where it quickly eclipsed such rivals as the lily and the carnation.

The initial impetus for the Dutch enthusiasm for tulips was provided by the flood of refugees and immigrants who poured across the borders of the United Provinces from the southern Netherlands at intervals throughout the Dutch revolt. Tens of thousands of Protestants living in the Spanish lands fled north in order to keep their religion and escape intermittent bouts of persecution. In some instances the influx of immigrants more than doubled the size of Dutch towns: 28,000 refugees arrived in Leiden between 1581 and 1621, and the total population quadrupled from 12,000 to 45,000, while in Amsterdam, throughout the seventeenth century, the majority of men marrying within the city walls had not been born there. The immigrants were willing to work hard and they often had capital to invest, substantially adding to the sum total of Dutch prosperity. The majority were capable artisans who could contribute useful skills – the foundation of the famous Amsterdam diamond trade was directly attributable to immigrants from Antwerp – but among their numbers were many of the wealthiest merchants of great towns such as Brussels and Antwerp. These men included a number of early tulip enthusiasts who brought their bulbs with them, introducing many varieties new to the United Provinces. By swelling the number of bulbs in cultivation, the refugees must have also made the flower significantly more widely available than it had once been.

The tulip was not just popular among immigrants; many Dutchmen were also becoming passionate about the flower. Tulips began to be cultivated throughout the Republic, where they were admired by an increasing number of knowledgeable connoisseurs and grown in a profusion of varieties from Rotterdam in the south of the country to Groningen in the north. In the United Provinces, unlike the rest of Europe, tulip connoisseurs were rarely aristocrats.

They were, rather, members of the new ruling class of the Republic: a group of rich and influential private citizens whom the Dutch called 'regents'.

The regents of a Dutch city typically included particularly well-to-do second- or third-generation businessmen, some lawyers and perhaps a physician or two. As a rule they were wealthy enough to live by investing their money in bonds, foreign trade or, closer to home, one of the many profitable schemes for reclaiming land from the sea or draining lakes and marshes to create new farmland. They were thus freed from the day-to-day cares of earning a living and formed a self-perpetuating ruling class whose members filled the principal posts in the provincial parliaments and town councils.

The few Dutch connoisseurs who were not regents were merchants, some of whom were at least as wealthy as their compatriots, but who nevertheless earned their living by taking an active part in the running of some business or other. The men of this class were generally accorded an honorific title which recognized their particular calling – so that a man named, for example, de Jonge who was involved in the fisheries would be known as 'Seigneur de Jonge in Herring' – and they tended to reinvest the profits they made in their own businesses. They had less time for their gardens than did the regents, but, even so, a number of the richest merchants did become noted tulip lovers.

The flower was, in fact, perfectly suited to the United Provinces. It was not only fashionable and far more intricately coloured than other garden plants; it was also unusually hardy, which meant that novices as well as expert horticulturalists could grow it successfully. The bulbs flourished best in poor, sandy soils of the sort found in several parts of the Republic, and particularly in Holland, where a belt of dry, white earth ran parallel with the coast all the way from Leiden up to the city of Haarlem, just to the west of Amsterdam, and then on to Alkmaar at the northern tip of the province.

What mattered most, however, was the tulip's new status as a symbol of wealth and good taste. Beginning in about 1590, the United Provinces became, quite unexpectedly, by far the richest country in Europe. For more than half a century, enormous sums of money poured into the country, greatly expanding the ranks of wealthy merchants. These men could afford to spend lavishly on things of beauty.

A number of contemporary writers have preserved the names of some of the Dutch connoisseurs who collected tulips in the first decades of the seventeenth century. They include several of the richest and most influential people in Holland – men such as Paulus van Beresteyn of Haarlem, who was once a regent of the local leper-house, and who grew tulips within the city walls, and Jacques de Gheyn, a painter from The Hague. De Gheyn was a well-known patrician and an acquaintance of Clusius's who was sufficiently passionate about gardening to complete a volume of flower paintings, twenty-two pages long, which he sold to the Holy Roman Emperor Rudolf II. He was one of the few connoisseurs whose real wealth is known with some accuracy, since he had his capital formally assessed in 1627, two years before his death. This audit showed he was then worth no less than 40,000 guilders.

Another tulipophile whose name figures in old records was Guillelmo Bartolotti van de Heuvel – who was actually

thoroughly Dutch, and owed his bizarre name to the fact
he had been adopted by a childless uncle from Bologna.
One of the two richest men in all Amsterdam, with assets
worth a staggering 400,000 guilders in total, he was quite
probably the wealthiest private individual ever to participate
in the tulip trade. Having built his fortune in business,
Bartolotti could afford to devote his leisure time to cultivating a celebrated garden right in the centre of Amsterdam.
From the scant descriptions which have survived of it, it
appears it was laid out to a highly symmetrical and fiercely
formal plan. Almost certainly it would have been the garden
of a true connoisseur, following the contemporary fashion
for flowers to be planted one to a bed so they could be
admired in splendid isolation.

The vast influx of wealth that made a rich man of
Guillelmo Bartolotti was primarily a consequence of the
Dutch revolt. In the previous century, the Republic's largest
town, Amsterdam, had been a city of only modest import-
ance, while Antwerp, in the southern part of the Netherlands,
was both the largest port and the wealthiest town in Europe.
Huge quantities of goods from the Baltic, Spain and the
Americas passed through the city on their way to the Holy
Roman Empire and the other states of northern Europe.
But with the seizure of Flushing by the Sea Beggars in the
first days of the rebellion, the Dutch were able to cut off
much of the city's commerce by blocking the mouth of the
River Scheldt, which gave Antwerp its access to the sea.
The blockade was a catastrophe for the Flemish town. Much
of its considerable trade was diverted north to Holland,
where Amsterdam became the principal beneficiary.

At about the same time, the Dutch broke what had

previously been a Spanish monopoly by opening trading links with the East Indies. To Europeans of the seventeenth century, the Indies were a source of almost unimaginable wealth. They overflowed with luxury goods, from spices to Chinese porcelain, which could not be obtained elsewhere. These goods could be purchased relatively cheaply in the East and were hardly bulky, yet they could earn a fortune at home. A single cargo of spices was worth many times more than the same tonnage of timber, grain and salt – the commodities on which the Netherlands had long depended – and could be turned into spectacular profits if it could be brought safely home. Dutch merchants were quick to recognize the potential of trading with the East. By 1610 they had established outposts on a number of Indonesian islands, and despite the constant threat of Spanish attack, fleets laden with peppercorns and nutmeg, cinnamon, cloves, sugar, silks and dyestuffs were sailing regularly to the United Provinces. The merchants of Amsterdam called these new commodities the 'rich trades', and with good reason – a single voyage to the Indies could yield profits of anything up to 400 per cent.

The surplus of wealth that now surged into the Republic touched the lives of thousands of Netherlanders. By 1631, fully five-sixths of Amsterdam's richest three hundred citizens had a stake in the rich trades, and both the Dutch merchant class and the regents who backed them and invested in their enterprises were enormously better off, on average, than their contemporaries in England, France or the Empire.

By the standards of the time, the most successful Dutch merchants were astonishingly wealthy. In the first half of the seventeenth century, a trader of the middle rank might have thought himself comfortable if his income reached 1500 guilders a year, and well-off if it topped 3000, while those below him in the social scale – clerks, shop owners and others with some claim to the title 'gentleman' – earned an average of a third or a fifth as much: perhaps 500 to 1000 guilders a year. But for the men such as Bartolotti, who had substantial stakes in the rich trades, incomes of 10,000, 20,000, even 30,000 guilders a year were possible. The richest of the lot was Jacob Poppen, the son of a German immigrant who had built his fortune trading with the Indies and with Russia. He was worth 500,000 guilders when he died in 1624. Adriaen Pauw, a regent who became a pensionary of Holland* and eventually one of the most prominent politicians of the United Provinces, amassed a fortune of 350,000 guilders from his successful investments, and by the 1630s another ten Amsterdammers possessed 300,000 guilders or more.

Today, men of comparable wealth dress in the finest clothes and travel by private jet and limousine. But even at the height of the Dutch Golden Age, visitors to the Republic found it hard to distinguish the wealthiest members of the regent and merchant classes from their countrymen. Even the richest of them dressed in clothes of the most severely unadorned variety, following the national fashion for large, wide-brimmed hats, tight trousers and a heavy jacket. Under that they sported a doublet, resembling a waistcoat, all in black, with substantial white ruffs at the throat and wrist, knee stockings and tight black shoes. Their wives and

---

* A senior government position equivalent, perhaps, to a modern-day Secretary of State.

daughters dressed in drab bodices and floor-length dresses, over which a lace apron often appeared. In winter, to keep out the pervasive Low Countries chill, men and women alike donned elegant fur-lined dressing gowns which were worn over all the other clothing in the home and at the place of work, but otherwise it was customary to avoid any sort of display of wealth. Women rarely even displayed their hair, preferring to hide it under a tight white cap, and though Dutch men did style theirs in something approaching Cavalier fashion – long and curled at the shoulders, with moustaches and a small triangle of neatly trimmed beard – on the whole the national dress sense was resoundingly Puritan.

But however modest their dress, Dutch regents and merchants were not immune to the temptation to display their wealth. The riches which came in with the tides and flowed into the coffers of these merchants and their backers had to find outlets of some sort. Some of the money, spent on food and wine, or used to import produce to the towns from the countryside, trickled down to the lower levels of society and helped to raise standards of living throughout the Republic. Much was saved or reinvested. Still, there is no question that the profits of the rich trades also fuelled consumption of all manner of luxuries, from great houses to paintings to tulips, making possible the remarkable variety and richness of the Golden Age that the United Provinces enjoyed between 1600 and 1670.

It was a time of tremendous cultural progress. The arts flourished as never before, fuelled not only by the establishment of Leiden and other universities and schools, but also by the arrival of many painters and writers from the south. So many artists, indeed, were looking for work that it became possible to commission new paintings or a play for a fraction of the usual cost. Many towns and private citizens took advantage of this fact, and visitors to the United Provinces were always greatly impressed by the variety and magnificence of the canvases, the tapestries and the statuary that turned up in the most surprising places. At the same time, several of the most brilliant artists were developing new techniques of realistic portraiture, creating the styles which men of the stature of Rembrandt (a Leiden miller's son) and Frans Hals (a refugee from Antwerp) perfected. Architecture, too, enjoyed a renaissance as the new republic commissioned many imposing public buildings, and there were more books, more pamphlets and more schools.

Individual Dutchmen also acquired a taste for building work. One of the principal reasons for the ever-increasing popularity of the tulip was the new-found passion among Dutch merchants and gentry for building grand country houses where they could enjoy – and indeed show off – their burgeoning wealth. Substantial mansions sprang up in clusters outside the richest Dutch towns: at Leiderdorp, a village on the outskirts of Leiden, among the rolling sand dunes on the coast west of Haarlem, and on the River Vecht where it flowed from Utrecht to Amsterdam. They were typically built in the Classical style, fully staffed, amply proportioned, and set in extensive grounds that generally included formal gardens as well as parkland. For busy and successful merchants and for the hard-working members of the regent class, they acted as retreats from the hasty world of the city.

Social historians have found in this passion for house-building an indicator of changing moods among the ruling classes of the United Provinces. During their Golden Age, the once sober, God-fearing Dutch – so Calvinist that their society frowned on ostentation in all its forms and ministers were fined for venturing the merest semblance of a joke in church – slowly acquired something of a taste for display. From this point of view, perhaps the most interesting product of the building craze was Zorghvliet ('Fly from Care'), the country home of a prominent regent named Jacob Cats.

Cats was a mild-mannered and extremely religious man who pursued dual careers as a politician and writer. He became without question the most widely respected Dutchman of the age. His fortune rested on his immense success as the author of popular moralistic verse, which sold in staggering quantities throughout the Republic. A typical Cats stanza, in which the poet rather savoured the opportunity to warn a beautiful young girl not to trade on her good looks, went like this:

> Blonde turns to gray
> Light-hearted becomes grave
> Red lips will turn blue
> Beauteous cheeks will be dull
> Agile legs become stiff
> And nimble feet halt
> Plump bodies lean
> Fine skin wrinkled

Father Cats, as he was universally known, turned out more

than a dozen books filled with this sort of verse, and something like 50,000 copies of his complete poems found their way into Dutch homes; often a volume of Cats would be the only book in the house apart from a Bible. Many Dutch families regarded him fondly as an honest source of wisdom and saw his verses as a reliable guide to the moral problems of the day. If Jacob the poet thought there was nothing wrong about owning a country retreat, it was difficult to argue that there was.

This fashion for sumptuous country houses led naturally to the planting of many extensive country gardens. Dutch interest in horticulture had begun to flourish in the previous century and still showed no signs of abating. The grounds of the Lord Offerbeake's residence in Alphen, near Leiden (which the English Member of Parliament Sir William Brereton visited in 1634), contained 'spacious gardens and mighty great orchards, and a store of fish-ponds', as well as twelve different varieties of hedgerow, a maze, long wooded walks, and of course a good number of flower beds. To be sure, Offerbeake's was one of the grander plots in the United Provinces, but other wealthy men followed his example as best they could. Such gardens were regarded not so much as places of relaxation but as a means of displaying the proprietor's collection of plants.

Despite the censure the tulip occasionally attracted from the more Calvinist elements of Dutch society, the tacit approval of moralists such as Father Cats encouraged the connoisseurs' enthusiasm for the tulip, whose beauty was after all one of those minor miracles wrought by God, and whose cultivation involved honest toil in the open air (an activity heartily recommended by Cats himself). The flower quickly became a prominent feature of many of the grandest new residences. One tulip garden that we know something about was planted at a country house called the Moufe-schans, which was celebrated in an epic poem of some 16,000 verses published in 1621 by a virulently anti-Spanish minister called Petrus Hondius. The Moufe-schans, which was built on the site of some German entrenchments dug during the Dutch revolt, and whose decidedly unbucolic name actually means 'Trenches of the Krauts', was owned by Johan Serlippens, the burgomaster of Terneuzen. Serlippens invited his friend Hondius to stay with him, and in time the clergyman planted a herbal garden in the grounds which included six full beds of tulips, an impressive quantity for the time. Hondius probably had some of his bulbs from Clusius, and others from an apothecary friend, Christiaan Porret of Leiden.

Hondius was no tulip maniac. He grew all manner of plants in Serlippens's garden, from carnations to hyacinths and narcissi, and he looked down on those who favoured the tulip over other flowers, writing scathingly in his verse of those who had allowed themselves to get too caught up in the burgeoning craze:

> All these fools want is tulip bulbs
> Heads and hearts have but one wish
> Let's try and eat them; it will make us laugh
> To taste how bitter is that dish

But the poet was himself far from immune to the new flower's allure. In *Of de Moufe-schans* he challenged contemporary painters to capture the tulip's beauty on canvas – only to admit, a line or two later, that the task was quite impossible. In his garden alone, Hondius wrote, the tulips on display exhibited a greater profusion of colours than artists even knew existed. The success of his epic poem – a treasure-house for social historians which contains copious information concerning not only gardening but also the lives and habits of the country people of the time – brought some of the most eminent men of the day to Serlippens's house; the visitors we know about included Maurice of Nassau, the new prince of Orange and commander-in-chief of the Dutch armies fighting the Spaniards, who was one of the most celebrated soldiers of the day. Maurice must have liked what he saw of Hondius's garden, for tulips were henceforth grown in the grounds of his palace at The Hague, in such quantities that they were eventually offered for public sale. (Sir William Brereton, who visited the palace a decade or so later, was able to purchase a hundred bulbs there for the modest price of 5 guilders.)

By 1620, then, the tulip was an established favourite with many of the Dutch elite and the private passion of some of the most influential men in the Republic. But, as the example of Prince Maurice shows, it was not yet so widespread as to be a commonplace for every citizen of the United Provinces. The flower was, even now, still comparatively rare, and some of the most highly sought-after varieties were hard to obtain at any price. It was only in the coming decade that this scarcity would be properly addressed.

CHAPTER 7

The Tulip in the Mirror

Other regents had their country houses. Adriaen Pauw, the immensely wealthy pensionary of Holland, owned a castle.

It was only a ruin, but it stood at the centre of a substantial estate called Heemstede, which Pauw acquired in 1620, and occupied the only high ground between the North Sea coast and Amsterdam. From the top of the crumbling walls, Pauw enjoyed commanding views over the heartland of the Dutch Republic. On clear days, he could see all the way to the roofscape of Amsterdam. Even when the weather was overcast, his country home offered an arresting view of bodies swinging from the gibbets that stood outside the walls of Haarlem, less than a mile to the north.

Heemstede became Pauw's greatest indulgence. The pensionary lavished money on his estate. He tore the old castle down and replaced it with a modern manor house where he entertained not only the most important men of the Republic but, on separate occasions, the Queen of England and the Queen of France. The interiors were suitably splendid; Pauw packed his new home with expensive furniture, finely woven tapestries and the best paintings. There was a trophy room full of burnished armour and a library of 16,000 books, a truly gigantic quantity for the time.

While the manor house was being built, Pauw set about improving the grounds. Always an enthusiastic investor in land-reclamation projects, he had tons of poor topsoil scraped away to reveal more fertile earth beneath. He encouraged farming and even light industry in the remoter portions of the estate, enlarging the population of Heemstede to more than one thousand people over time.

But Adriaen Pauw's greatest joy was not his manor house but his garden. It was carefully laid out just in front of the house after the formal fashion of the time, with long tree-shaded walks bisecting ornamental lawns and flower beds filled with roses, lilies and carnations whose colour splashes set off the geometric precision of Pauw's box hedges and paths to perfection. And, in a prominent position in the centre of the garden, the new lord of Heemstede planted a single bed of tulips.

There was one very peculiar thing about Pauw's garden, though. It wasn't something that visitors noticed straight away; in fact, although the burgomaster entertained lavishly and even permitted sightseers to wander around the estate on weekdays when he was occupied in Amsterdam, some left without ever realizing it was there. But that was not altogether surprising. It was not something Pauw wanted his visitors to see.

The secret of the gardens of Heemstede was a weird contrivance of wood and cunningly angled mirrors which stood in the middle of the tulip bed. It was a looking-glass cabinet, designed to multiply whatever stood before it. Its purpose was to create an illusion of plenty where really there was none.

From a distance, and with this strange invention's help, Pauw's single tulip bed looked densely planted with hundreds of brilliant flowers. It was only when a curious or appreciative visitor approached more closely that he would realize this was just an optical illusion. The mirrors of the wooden cabinet had turned the few dozen tulips in Pauw's collection into a profusion.

For the lord of Heemstede, the looking-glass cabinet was an admission that there were some things even he could not buy. Rich and influential though he was, the pensionary of Holland could not obtain enough tulips to fill his garden, and all the efforts of the best gardeners in the province could not persuade the bulbs he already had to multiply as quickly as he wanted.

Pauw's problem was a simple one. The superbly fine varieties that he collected were extremely rare, because they were the products of a long process of selective breeding. Ever since the earliest Dutch tulips had flowered in Walich Ziwertsz.'s Amsterdam garden, connoisseurs had been carefully selecting the most exquisite specimens, cultivating them with special attention, and crossing them with other fine bulbs to create ever more beautiful varieties. Thus, while the earliest, rudest tulips had had decades to multiply, the most valuable flowers, those which bore the most striking colours, were only recent creations. The finest tulips of all were available in such small quantities that even Adriaen Pauw could not obtain them.

Of all the varieties acclaimed 'superbly fine', easily the

most coveted was a flower called Semper Augustus, which was the most celebrated, the scarcest and, by common consent, the most wonderful tulip grown anywhere in the United Provinces during the seventeenth century – and thus also by far the most expensive. Semper Augustus was a *Rosen* tulip, but to call it simply a red and white flower would be like describing rubies and emeralds as red and green stones. Everyone who saw it concurred that it was a plant of quite exceptional beauty. It had a slender stem which carried its flower well clear of its leaves and showed off its vivid colours to the best effect. Beginning as a solid blue where the stem met the flower's base, the corolla quickly turned pure white. Slim blood-coloured flares shot up the centre of all six petals, and flakes and flashes of the same rich shade adorned the flower's edges. Those fortunate enough to see a specimen of Semper Augustus in flower thought it a living wonder, seductive as Aphrodite.

In truth, though, very few people ever had this privilege. Although the tulip was endlessly hymned by the connoisseurs, illustrated in more tulip books than any other variety, and mentioned so frequently in connection with the bulb craze that it has become virtually synonymous with it, Semper Augustus was practically never actually traded. It was so rare that there were simply no bulbs to be had.

It is with the mysterious Semper Augustus, then, that the first symptoms of what would become known as tulip mania appear. Quite how the flower first made its way to the United Provinces is not known. According to the Dutch chronicler Nicolaes van Wassenaer, who is virtually the only reliable source on the subject, the tulip was grown originally from seed owned by a florist in northern France, but, not recognizing its value, he disposed of it for a pittance. This must have been around the year 1614. When the variety had eclipsed all other flowers, ten or twelve years later, connoisseurs from Holland hurried south to scour the nurseries and gardens of Flanders, Brabant and northern France for other specimens of the Semper Augustus. It was a difficult task, and they met with no success. A few similarly patterned flowers were located – one even received the name Parem Augusto, in honour of its apparent kinship – but somehow none quite matched the emperor of tulips in the vividness of its colour or the purity of its form.

This failure forced Dutch connoisseurs to try a new tack, and for a while they attempted to promote the most gorgeous specimens from their own collections as rivals to Semper Augustus. Van Wassenaer mentions the varieties Testament Clusii, Testament Coornhert, Motarum van Chasteleyn and Jufferkens van Marten de Fort in this connection, but striking though they were, none of these flowers excited anything like the admiration reserved for the red-flamed emperor. Nor did persistent rumours that a tulip eclipsing even Semper Augustus in beauty had been found growing in a garden in Cologne ever amount to anything.

The earliest mentions of the flower date only to the 1620s. By 1624 – according to van Wassenaer – no more than a dozen examples were in existence, and all twelve were in the hands of a single man who was generally rumoured to live in Amsterdam. His identity is one of the great mysteries of the tulip craze. Van Wassenaer is careful not to name him, and in the absence of any other evidence, it seems unlikely that the puzzle will ever be solved. This was perhaps what the reclusive connoisseur would have wished, because the chronicler makes it clear he was determined not to part with his flowers at any price.

He could have sold his bulbs easily. At a time when tulips were increasingly widely cultivated, the fact that no more than a dozen specimens of a superbly fine variety were in existence made them a fantastic rarity, and the evidence suggests the owner could have charged almost any price he cared to ask for a single bulb of Semper Augustus. Instead, he turned down every request to sell his flowers.

Throughout the 1620s, the tulip enthusiasts of the Republic bombarded the man with ever more extravagant offers for a single bulb. The amounts they were willing to pay were not just large, they were spectacular: van Wassenaer records that in 1623 the sum of 12,000 guilders was not enough to procure ten bulbs, and states that the reclusive connoisseur who grew the flower valued the private satisfaction of gazing on the beauty of Semper Augustus over any potential profit. Yet his refusal to consider offers merely drove his desperate colleagues to raise their bids. Next summer offers of as much as 2000 or 3000 guilders a bulb were made – and just as summarily rejected.

In the end, though, all the mysterious owner's efforts to control the supply of Semper Augustus proved to be in vain. Van Wassenaer explains that once, early on, the connoisseur who had discovered the variety did agree to sell a single precious bulb, for the not inconsiderable sum of 1000 guilders. When the flower was lifted from its bed, he saw that it had grown two offsets from its base. This discovery mortified the owner, who might reasonably have demanded 3000 guilders, rather than 1000, for the tulip, but it was a piece of tremendous good fortune for the buyer, who had every incentive to sell one of the offsets to recoup his investment in the flower. He now held the makings of a valuable collection in his hand.

From this uncertain start, Semper Augustus did very gradually become available to the handful who could afford to pay for it. Bulbs of this most highly sought-after of flowers appear to have produced viable offsets only rarely – this was a characteristic of the most superbly fine tulips, perhaps because they were more heavily infected with the mosaic virus than were ruder varieties – and even a decade later, only a handful existed. Of course, the continued rarity of Semper Augustus did not stop connoisseurs from coveting the flower; in fact it merely fanned their ardour. That was as good a measure as any of the frenzy for bulbs that was now beginning to well up inside the Dutch Republic.

The scarcity of tulips in seventeenth-century Holland was central to the bulb craze. To a Dutchman of the Golden Age, the tulip was not the mundane and readily available flower that it is today. It was a brilliant newcomer, still full of the allure of the exotic East, and only obtainable in strictly limited quantities. Because the most superb varieties were scarce, they were coveted; because they were coveted, they were expensive. And because they were expensive, they became increasingly lucrative to grow.

A handful of tulip connoisseurs had always produced their own flowers and been keen and able horticulturalists in their own right. For example, the brothers Balthasar and Daniël de Neufville – a pair of rich linen merchants from Haarlem – bred two new varieties, one a *Rosen* and the other a *Violetten*, which they grew in the garden of a house just inside the city walls which they had named 'the Land of Promise' Most of their contemporaries, however, were less skilled, and by the late 1620s it was increasingly apparent that demand for tulips could no longer be met simply by the exchange of small quantities of bulbs among connoisseurs. New tulip enthusiasts, who had none of the skills needed to breed varieties of their own and few of the connections necessary to obtain bulbs in the traditional way, had begun to enter the market. Some possessed extensive gardens and wished to cultivate tulips of many different varieties. These newcomers were forced to look elsewhere for their supplies.

They turned to the handful of professional horticulturalists who had begun to cultivate the fashionable new flowers. This was a key development in the history of the tulip, for there can be no doubt that without the efforts of the professionals many fewer new varieties would have been developed. The total quantity of bulbs in circulation would have been less too, and the speed with which the tulip spread across the United Provinces would have been reduced.

By 1630, professional flower growers could be found in almost every town in the Dutch Republic. Few of them produced bulbs on any scale. The majority had only small gardens and many were forced to hire land from local taverns or even – in some cases – monasteries, in order to deal with the rising demand for their flowers. Most cultivated all manner of flowers, although a number, such as Henrik Pottebacker of Gouda – creator of the *Rosen* varieties Pottebacker gevlamt and Admirael Pottebacker – had begun to specialize in tulips. They were expert horticulturalists and, just as importantly, they had a keen eye for what was valuable and what would sell.

At the top of the market, Semper Augustus's closest rivals included Viceroy – a big, bold, purple-flamed flower generally recognized as the king of the *Violetten* – and, at the head of the *Bizarden*, a flower called Root en Gheel van Leyde ('Red and yellow of Leiden'). At the bottom, the cheapest and least coveted flowers were simple unicoloured blooms, their petals coloured yellow, red or white. These were the earliest of all Dutch tulips and consequently also the most common.

Gardeners such as Pottebacker had not sprung from nowhere. They had learned their skills from earlier, less commercially adroit growers who had existed in small numbers since the end of the sixteenth century and scratched a living in the limited markets of the day. Clusius and his circle of aristocratic friends possessed a low opinion of these first professionals, criticizing them for their often alarming ignorance of botany and despising their willingness to bestow crudely populist names on the new varieties which occasionally appeared – more, perhaps, by luck than judgement – in their gardens. Nevertheless, they grew tulips, and they were learning.

By the beginning of the seventeenth century, the handful of pioneer bulb farmers (who then based themselves in the countryside just outside Brussels) were having to compete with an even more discreditable group of itinerant flower collectors. These restless individualists scoured the countryside – mostly in France – for unusual specimens and sold them to collectors, mostly in the Netherlands. They called themselves rhizotomi (the word is Greek for 'root cutters'), and even Clusius, in his declining years, found them useful sources of flowers he was no longer mobile enough to collect for himself.

At least a few of the rhizotomi were honourable men – Clusius names Nicolas le Quilt, of Paris, and Guilielmus Boëlius as reliable suppliers of the rare bulbs he still sought – but on the whole the collectors had a slightly unsavoury reputation. This was because it was easy for them to palm off ordinary seeds and bulbs as rarities, and claim substantial payments for their labours, safe in the knowledge that they would be long gone and back over the border in France before the flowers bloomed and the fraud was discovered. Since it was quite impossible for even an expert botanist such as Clusius to tell what sort of tulip would grow from an anonymous brown bulb, this problem was destined to cause all sorts of disputes.

The rhizotomi were not the only people searching the countryside for rare plants in the first years of the new century; wild tulips were increasingly available from apothecaries too, collected during trips to gather medicinal plants and herbs. Among those known to have stocked the bulbs are three Dutchmen: Willem van de Kemp of Utrecht, Petrus Garret of Amsterdam, and Christiaan Porret of Leiden.

Apothecaries were as commonplace in the seventeenth century as chemists are today. These early pharmacists peddled folk and quack remedies to those who could not afford the services of the few qualified physicians of the time. They wore the same uniform as doctors – black robes and coat, collar bands and a pointed hat – but their premises were easily identified by their traditional symbol, a stuffed crocodile, which generally hung suspended from the ceiling over the counter.

Some undoubtedly were men of integrity, but in the first years of the seventeenth century, apothecaries shared with the rhizotomi a certain notoriety as disreputable opportunists. They had only recently left the grocers' guild, to which they had belonged for centuries, to join the physicians – so recently, indeed, that their shops were still the only places where Dutchmen were allowed to buy fruit tarts. But many apothecaries had come up with better ways of making money than that. Their premises often doubled as clandestine drinking dens, and many secretly offered medical consultations, which were supposed to be the monopoly of physicians. Such men were happy to meet the growing demand for tulips by supplying dried bulbs. Some of their customers were genuine flower lovers and connoisseurs, but it appears that the more unscrupulous apothecaries also promoted bulbs as an aphrodisiac.

It was only gradually, between the years 1600 and 1630, that the buccaneering rhizotomi and the apothecaries were supplanted by the new breed of respectable professional nurserymen. Many of these growers were based in Haarlem, the second largest city in the province of Holland, which had been built on the sort of poor, sandy soil that was ideally suited to the cultivation of tulips. They favoured small plots of rented land just outside the city walls and within easy walking distance of the gates. According to Haarlem tradition, most of the city's tulip gardens were just outside the Grote Houtpoort ('the Great Wood-Gate') which guarded one of the two southern entries to the town. But perhaps the best of Haarlem's little flower farms were situated along the bosky Kleine Houtweg ('Little Wood-Road'), which ran from the other gate on the south side of the city down through an area still known today as the rose district and on to Haarlem Wood, which was the city's favourite beauty spot. More than twenty nurserymen were based along this road, and it was here that the tulip grower David de Mildt, who figures prominently in many of the surviving records of the mania, had his garden at a spot named Twijnderslaan. When de Mildt died, aged only thirty-three, at the height of the craze, his plot was taken over by another prominent tulip grower, Barent Cardoes, and renamed the Garden of Flora. Under Cardoes's management, it became one of the best-regarded bulb farms in Holland.

Barent Cardoes learned his trade working for another Haarlem grower, Pieter Bol. Bol was the creator of the *Violetten* Anvers Bol and several other superbly fine varieties, and possibly the richest tulip grower of the age. Unlike the majority of his fellow growers, it would appear he was a patrician and a connoisseur who employed professional gardeners such as Cardoes to do much of the actual work involved in cultivating his bulbs. But some distance south of the city, in the lordship of Vianen, lived another grower who came from a much humbler background. His name was Francisco Gomes da Costa, and he was probably the most industrious of all the horticulturalists in the United Provinces.

Da Costa was a Portuguese grower who made a name for himself with the sheer variety of tulips he created. He seems never to have mastered the Dutch tongue with much confidence – the manuscript of a gardening book he commissioned for his own use lists the names of all his flowers phonetically for his benefit – but he was an unparalleled innovator in the garden. No fewer than eight varieties bore his name, the largest number of varieties named after one man. One of his most famous creations was Paragon da Costa. A Paragon was generally defined as a variety that was an improvement on an existing flower, usually because its colours were finer or more intense. On that basis, Francisco da Costa's proudest achievement was probably Paragon Viceroy da Costa, a tulip which claimed to improve on the unimprovable Viceroy.

For an immigrant such as da Costa, tulip farming was attractive for precisely the same reasons that it appealed to many of the Dutch. Little investment was required, a small plot of land and some tulip seed or bulbs being all that was required to start; tulips were hardy and grew well in poor soil; and bulb growers were not required to belong to any of the restrictive and expensive guilds which so rigorously controlled most of the trades and professions in the Dutch Republic.

To anyone who was at all horticulturally inclined, however, it was the profits to be made in the tulip trade which proved most alluring. Individual growers certainly became wealthy. Pieter Bol's name is mentioned as one of those who profited most from the flower business, but when the Haarlem dealer Jan van Damme died in 1643, he left an estate consisting primarily of tulip bulbs which was valued at 42,000 guilders, a fortune that ranked him alongside many of the wealthy merchants who had made their money in the rich trades.

Where did all this money come from? Successful growers such as van Damme owed their success to an ability to exploit every possible market for their bulbs. Most found ready customers among the connoisseurs and the owners of fashionable new country houses, but they were happy to sell their bulbs to members of the emerging merchant class as well. And, from as early as 1610, a few intelligent horticulturalists were selling their tulips in the Holy Roman Empire and, no doubt, the southern Netherlands and northern France. What started as a very minor export trade indeed grew, slowly but surely, to the point where in the first quarter of the eighteenth century the Dutch were shipping cargoes of bulbs to North America, the Mediterranean and even the Ottoman Empire.

Perhaps the first Dutch bulb dealer to move into the export trade was Emanuel Sweerts, another old friend of Clusius's who kept a curiosity shop in Amsterdam and was active in the first decade of the century. He not only imported bulbs from all over Europe, but began to offer tulips for sale at the *Messe*, a vast fair held each year at Frankfurt-am-Main. (The Frankfurt Book Fair, which attracts tens of thousands of publishers to the city each year, is actually a successor to this tremendous medieval market.)

The increasing professionalism of the bulb trade posed one important problem for men such as Emanuel Sweerts. Tulips were in flower for only a few days each year; they had to be sold as bulbs. But these plain brown packages offered no clue to the glories they concealed within, and certainly did not look an enticing investment. Sweerts came up with the solution: a catalogue packed with illustrations portraying his tulips in full bloom. He persuaded the most eminent of all his clients, the Holy Roman Emperor Rudolf II, to pay the print bill – the same emperor who had once dismissed Clusius from the imperial service now dabbled in tulips in between conducting the alchemical experiments that were his chief passion – and published his catalogue, the *Florilegium*, in Frankfurt just before his death in 1612. The *Florilegium* was modelled on contemporary herbals; there was very little text, but each of the tulips within it was accorded a concise description, in Latin, which gave the essential information about its shape and colour.

Only two years after the *Florilegium* first appeared, a Dutch artist called Chrispijn van de Passe produced a similar book called the *Hortus Floridus*. Van de Passe, the son of a Flemish engraver, was only seventeen years old when his book appeared, but it proved to be one of the most successful botanical works of the age and it was quickly translated from the original Latin into French, English and Dutch. The Dutch edition featured a list of the leading tulip enthusiasts of the early seventeenth century, and later impressions included an appendix devoted entirely to the flower which shows that a lively trade in bulbs already existed between the United Provinces and Germany.

It was not long before the *Hortus Floridus* began to serve as a handy catalogue for nurserymen who did not enjoy the luxury of a wealthy patron willing to bankroll the production of a book such as the *Florilegium*, which was devoted to cataloguing the stock of a single garden. But there was a limit to the usefulness of a general printed catalogue, particularly in the early days of the tulip trade, when every grower offered his own unique varieties for sale. This problem was solved by the introduction of one of the most notable traditions of the bulb craze: the production of richly illustrated, privately commissioned manuscripts called tulip

books. A significant number of these albums were produced
for individual Dutch horticulturalists and nearly fifty are
known today; they were anything up to 500 pages long and
typically contained one illustration per page, executed in
watercolour or gouache. Each picture was generally accompanied by the tulip's name, but only very occasionally by
any information regarding price. There is the suspicion
that bulb growers, like many modern-day antique dealers,
preferred to price their bulbs according to their assessment
of their customers' wealth.

Customers who paid more than they anticipated for their
bulbs were not the only people to be short-changed by the
owners of the tulip books; the artists who illustrated them –
a few of whom were eminent painters – were generally very
poorly paid for their efforts, perhaps earning only a handful
of stuivers per page. Marginal notes in a book painted largely
by Jacob van Swanenburch of Leiden, the master who taught
Rembrandt van Rijn, show that the painter completed 122
flower pictures for a fee of just over 6 stuivers per painting.

Jacob van Swanenburch was not the only highly regarded
artist to contribute to a grower's tulip book. Judith Leyster,
the only woman who actually earned her living as a painter
in the United Provinces during the Golden Age, painted
two *Rosen* tulips for an album now commonly known as
'Judith Leyster's Tulip Book' in her honour, although the
rest of the paintings are by other hands; and Pieter Holsteijn
the Younger illustrated a manuscript for a grower named
Cos which carries the date 1637 and, unusually, gives not
only the flowers' names (some of them in the form of a
riddle or a rebus) but their price and the weight of each
bulb when planted. It consisted of fifty-three gouaches of

tulips, with twelve additional drawings and some water-colours of carnations.

A close study of this and other flower albums shows that many of the artists who produced them created something approaching a production line of illustrations by arranging for their assistants to paint the flowers' leaves and stems (often in a hackneyed style which probably bore at best a sketchy resemblance to their real appearance) and executing only the difficult portion – the petals – themselves. Others copied sketches of the rarest varieties from earlier books, although some of the tulips were so scarce they must have been included more for the sake of completeness than anything.

To Dutch nurserymen, the tulip books were a valuable sales tool which could be used to attract more customers and lure existing ones to try new varieties. But the surviving albums, thronged with page after page of all but identical *Rosen*, *Violetten* and *Bizarden* flowers, inadvertently make an important point about the often chaotic workings of the seventeenth-century flower trade.

One of the chief difficulties facing both growers and connoisseurs was the problem of distinguishing between strikingly similar varieties. Even the most knowledgeable dealers and growers must have found it difficult, if not impossible, to tell one *Rosen* tulip from another with almost identical markings, even though those varieties might be worth very different sums. This problem lay at the root of a number of the sometimes bitter disputes between growers and their customers which dot the surviving records of the flower trade.

The fact that tulips of the same variety differed one from another and from generation to generation did not help, and nor did the plethora of confusingly similar names that were bestowed upon new flowers by their creators. Outsiders found the chaotic nomenclature of Dutch tulips almost impossible to grapple with. In this early period there were no firm rules and certainly no central authority which could impose any sort of order on the way in which tulips were named. Anyone who created a new variety had the privilege of conferring a title upon it, and generally they chose either to give it an overblown name which hinted at the exceptional qualities they felt it possessed, or to name the flower after themselves. Quite frequently they managed to do both.

The man who started this craze was the bailiff of Kennemerland, the coastal region between Haarlem and the sea. He created a *Rosen* tulip of exceptional beauty and, casting around for a name to convey its excellence, he decided to christen it Admirael ('admiral'). Before long the Admirael name had become the highest epithet a tulip could aspire to, and other growers flocked to apply it to their own creations: Admirael Liefkens, Admirael Krijntje, Admirael van Enkhuizen and the most celebrated of all, Admirael van der Eijck. Foreigners sometimes made the mistake of believing that these flowers were named after naval heroes of the Dutch revolt but, of course, they really commemorated not sailors but the horticulturalists who had created the flower. At the time of the tulip mania there were already about fifty different varieties with the Admirael prefix, and another thirty or so which bore the rival title Generael ('general'). The Generaels included one flower which had been named Generael van der Eijck, perhaps in the hope of persuading potential buyers that its qualities matched those of the Admirael tulip.

Once the fashion for Admiraels and Generaels had run its course, growers took the logical next step of searching for new superlatives and created a class of plants named Generalissimo. Next came varieties named after real classical heroes such as Alexander the Great and Scipio, and eventually two tulips from Gouda titled, with breathtaking arrogance, 'Admiral of Admirals' and 'General of Generals'. At least these really were superbly fine varieties, noted for their size and fiery scarlet stripes.

Such practices meant that many inferior tulips received the Admirael or Generael names, and customers could not necessarily even determine the sort of flower they were buying simply from its title. Generaels, for example, were almost always *Rosen* tulips, but at least three *Violettens* bore the name, and there were *Violetten* and even *Bizarden* Admiraels. Naturally all this confusion meant that growers had to do what they could to publicize the new varieties they had created. One contemporary writer explained how this was done:

> If a change in a Tulip is effected, one goes to a florist and tells him, and it soon gets talked about. Everyone is anxious to see it. If it is a new flower, each one gives his opinion; one compares it to this, another to that, flower. If it looks like an Admirael you call it a Generael, or any other name you fancy, and stand a bottle of wine to your friends that they may remember to talk about it.

Talk they did. By 1633, the combined efforts of the

growers and the connoisseurs, the rhizotomi and the apothecaries, had all but solved the old problem of scarcity. Tulips
were at last widely available throughout the Netherlands. A
total of some five hundred different varieties were by now
being grown in the Dutch Republic alone – some superbly
fine and extremely rare; others, still beautiful, rather easier
to obtain. And as the supply of bulbs steadily increased, the
flower began to attract new admirers among the tradespeople
and workers of the Dutch Republic – men who had not
until now been able to afford tulips, or displayed much
interest in the bulb trade.

In part, this was the work of the growers. Their most
important customers, the connoisseurs, were demanding
ever finer and rarer tulips, and this left the bulb farmers
with the task of disposing of increasing quantities of the
older, less spectacular varieties which naturally made up the
bulk of their stock. They solved the problem by selling
these flowers at low prices to new clients who had heard
much excited talk about the beauty of the more fashionable
varieties and wanted tulips of their own. Some of the more
ambitious growers even took to offering unwanted bulbs to
members of the army of pedlars who travelled from town
to town selling their wares at local fairs and markets. These
men hawked the flowers far and wide. They helped to
introduce the ruder varieties to farmers, labourers and polder
boys out in the countryside and spread the gospel of the
tulip far and wide.

In greater measure, though, the interest that many Dutchmen now developed in the flower trade owed less to the
tulip's natural beauty than it did to the dawning realization
that there was money to be made in bulbs. That was

something worth investigating. For money, despite the enormous wealth now flowing into the Republic, was something many of its citizens saw all too little of.

CHAPTER 8

Florists

Foreigners who marvelled at the wealth the Dutch enjoyed during their Golden Age never ceased to wonder how they did it. The regents and great merchants of the United Provinces might be rich, but the country they lived in was one of the poorest places in Europe in terms of natural resources. Few other nations were quite so lacking in fertile land, charming countryside and a pleasant climate as the Dutch Republic; from the war-ravaged territories of the south to the peat bogs that sprawled across the northern provinces, there was almost nothing to suggest that this was a land of any promise.

Here was a nation described by one scornful Englishman as 'an universall quagmire ... the buttock of the world', a country whose greatest city – Amsterdam – had been built on a swamp and could only be reached by braving the Zuyder Zee, a 50-mile-long inland sea full of sandbanks and treacherous shallows. It was a place where the air, in the words of the English ambassador Sir William Temple, 'would be all Fog and Mist, if it were not clear'd by the sharpness of their Frosts', where the weather was 'violent and sur-prising', and which was so unhealthy, chill and damp that it seemed to breed fevers and plague.

For the regents of the Dutch Republic, money made this situation tolerable. Farmers (or at least those fortunate enough to till the scarce fertile land north of the Waal and the Maas) also did well during the Golden Age. There were many mouths to feed in the Republic, and there was additional demand for their produce from the Holy Roman Empire, where the Thirty Years' War between the Protestant north and the Catholic south raged between 1618 and 1648, devastating local agri-culture. But for ordinary workers – the weavers and car-penters, smiths, cobblers and market traders who lived in the towns and made up what the Dutch called the artisan class – life in the United Provinces could be very hard.

In the seventeenth century, almost all Dutch artisans worked long hours for low wages. When the day's work was done and they could finally go home, it was to cramped and sparsely furnished one- or two-room houses that were in such short supply that rents were high. Even the national diet was monotonous. To people trapped in an existence such as this, the idea that one could earn a good living by planting bulbs and sitting back to watch them grow must have been irresistible.

For many years, most artisans had begun their working day before dawn and finished it after dusk. By 1630, the clamour that arose from city workshops as they opened for business in the small hours of the morning was so great that several towns had been forced to pass decrees for-bidding fullers from beginning work before two in the morning, and hatters from starting any earlier than four.

Blacksmiths suffered the greatest restrictions: their smithies were so noisy that they remained closed by order until the bell which announced daybreak had been rung.

During these long days, Dutch artisans were sustained by nothing more than snacks of cheese and raw pickled herring and a dinner, taken in the middle of the day, which typically consisted of the national dish, a meat stew known as *hutspot* made of chopped mutton, parsnip, vinegar and prunes boiled in fat. A good *hutspot* was supposed to be left to simmer for at least three hours, but when times were bad and the work was hard it was often cooked for no more than an hour so that when served it was still – in the words of one appalled French visitor – 'nothing more than water full of salt or nutmeg, with sweetbreads and minced meat added, having not the slightest flavour of meat.'

For many of the Dutch, though, even a poor *hutspot* was at best an occasional luxury. Those who could not afford meat lived on vegetables and the sticky black rye bread of the time, which was sold in huge loaves weighing 12 pounds or more; in poorer households, the mother might buy a single one of these loaves to feed her whole family for a day. Even when other food was available, Dutch eating habits were generally very conservative. Seafood, for example, almost always meant either herring or cod; mussels, though available, were despised as the poorest sort of food, and the servants at one grand house were so disgusted at being asked to eat salmon that they begged their mistress to promise she would not serve it to them more than twice a week.

With dinner over, work began again immediately and lasted at least until dusk, and much later if it was possible to continue under artificial light. During the Golden Age, fourteen-hour days were thought perfectly normal, and at Leiden, in 1637, hard-pressed cloth workers who had just worked a sixteen-hour shift needed money so badly that they asked to be put on overtime. Nor was there much time off; everyone worked six days a week, and one of the less welcome results of the Reformation had been the abolition of a good number of holidays which had previously been celebrated on Catholic saints' days.

The artisans rarely complained about this because they were paid by the hour. The money they could earn thus varied according to the number of hours it was possible to labour in a week, so a job that generated a little surplus income in the summer might become one that paid not much better than starvation wages when the days closed in for winter. Even when times were good and the days were long, the pay in most jobs amounted to something between half a stuiver and two stuivers an hour, and hundreds of thousands of Dutch people worked long and hard for a guilder a day, or a little less. The upshot was that at a time when Sunday working was not permitted and a family of five needed a minimum income of 280 guilders a year simply to avoid starvation, a Dutch artisan in regular work often expected to earn an annual wage of no more than 300 guilders.

Those who did take home more than this were not necessarily that much better off. Most of the trades in which an artisan could hope to earn a decent living were still controlled by guilds, which imposed considerable dues and expected their members to contribute to the costs of the frequent banquets and receptions that marked the course of a guild's year. A good number of artisans who had successfully completed the lengthy and poorly paid apprenticeships never could afford to pay such sums and had to remain journeymen all their lives. Even at the height of the Golden Age, when wealth was pouring into regent coffers from investments and the rich trades, the master craftsmen of the Republic who had surmounted every obstacle and joined their chosen guild were generally so poor that they could not afford to hire apprentices of their own to help them.

From this perspective, it is obvious that although the United Provinces were rich, few of those who lived there could be considered wealthy. Some artisans did earn good livings, it is true, and even the poorer ones were paid something like twice as much as the poor of other countries. But taxes and prices were correspondingly high throughout the Republic. Those who did have jobs worried constantly about money, and their wives generally had to work to supplement the family income.

A typical Dutch family, then, had little money to spare, and would have owned relatively few possessions. If they were artisans and citizens of one of the great towns, as were more than a quarter of the population of the Republic, they probably lived behind a door made of oak, waxed or painted green, in one of the small, neat houses which lined the crowded streets. The interior was kept scrupulously clean – the Dutch fetish for cleanliness was something almost every traveller remarked on, and it was not at all uncommon for a house to be permanently damp from repeated scrubbing and for any visitors to be required to wear straw slippers over their outdoor shoes to keep out dirt. But it would also have been relatively bare. An artisan household might boast a table, a plain cupboard, some tableware and perhaps a few straight-backed chairs, which sold for a guilder apiece, but it took a long time to scrape together enough money to purchase the most expensive item of household furniture, a bed. The cheapest varieties, known as cupboard beds because they were set into a wall to help retain warmth, were so small that they required their occupants to sleep in a sitting position, and even these cost 10 or 15 guilders; only members of the merchant class could have afforded a modern free-standing bed at the enormous price of 100 guilders. Among the artisans, children slept on couches or boards, or in drawers under their parents' bed, and when they reached the age of fourteen they too were expected to find work and contribute what they could to the household.

By 1630, moreover, the precarious prosperity of the artisan class was increasingly under threat from the flood of Protestant refugees arriving from the south. Even in the previous century, the people of the United Provinces had begun to realize that their republic was becoming crowded, since most of the cultivable land, and thus much of the population, was concentrated in the three relatively fertile provinces which lay at the heart of the country: Holland, Gelderland and Utrecht. (One other reasonably prosperous area lay to the south, where the people of Zeeland mostly earned a living from the fisheries, but the remaining provinces were not capable of supporting many people.) With the arrival of tens of thousands of immigrants from the southern Netherlands, most of whom were seeking work, the population had swollen to some two million people. The fact that many of the southerners brought their wealth with them certainly helped ease the burden but, even so, overcrowding became a significant problem, and those who were not already wealthy could see that their chances of ever prospering were increasingly limited.

The United Provinces possessed, however, one vital national characteristic in greater measure than any other nation in Europe in the first half of the seventeenth century. This was the belief in the possibility of social mobility that was the birthright of every Dutchman. In France or the Holy Roman Empire, a peasant knew that whatever happened to him, he would always remain a peasant, just as a shopkeeper would be the son and the father of shopkeepers. But the United Provinces was a land where an immigrant's son had become the wealthiest man in the richest city on earth and been co-opted, despite his entirely humble origins, into the regent class; where the village labourer could try his luck in the towns, and where a moderately well-off artisan could and occasionally did invest his money by taking a minute share in a ship setting off to trade in the Baltic, reinvest the profit and work his way up until he himself became a shipowner. Opportunities, then, did exist, and people could see them and badly wanted to take them, but 'whenever there is a stuiver to be earned,' as the Flemish preacher Willem Baudartius put it in 1624, 'ten hands try to grab it.' If you were poor, and struggling to make a living in the over-supplied labour market of the Golden Age, you were actually more likely to slide down the social scale than to climb up it.

Still, for the Dutch, the Golden Age was an era pregnant with expectation of change. That emotion was felt by the poor at least as much as it was by the rich, and this did more than anything to persuade tradesmen and artisans to try their luck in the bulb trade. As demand for tulips grew, and the prices quoted for particular varieties increased year by year, it became increasingly obvious that there was money to be made here. From the early 1630s, then, a new sort of buyer began to nose about the nurseries of the Dutch Republic. The newcomers were not connoisseurs of flowers, and many of them knew little or nothing about cultivating bulbs. They called themselves 'florists' and they were interested only in making money out of tulips.

The first florists probably thought of establishing themselves as growers. The idea of taking a simple bulb and turning it into cash in the course of a single winter must have been a very attractive one, and naturally it appealed particularly to the itinerant, the indolent and the chancers in Dutch society – people with no fixed employment and no fixed income, who welcomed what seemed a fine opportunity to earn some easy money. Many honest artisans who worked enormously hard to make a fraction of what some tulip growers earned found the flower trade increasingly attractive as well. Equally naturally, tulips were less enticing to the better-off and those fixed in stable professions, who were already living a reasonably comfortable life.

The notion of creating a little tulip nursery would have come quite naturally to many of the florists. By the 1630s, the fashion for gardening, which had earlier been largely confined to the regent and merchant classes, had begun to spread much further afield. Many of the artisans who lived in cities such as Haarlem and Amsterdam had access to an allotment outside the city walls. Before the bulb craze got properly under way, these had mostly been used to grow root vegetables, but even then a handful could be surprisingly elaborate, as Sir William Brereton observed of a poor man's garden at Leiden: it contained topiary which 'portraited to the life in box all the postures of a soldier, and a captain on horseback'. Another English traveller, Peter Mundy, thought the pleasures of cultivating a little garden helped Amsterdammers to cope with the miseries of living in their marshy climate. 'The want of walking Feilds and Meddowes,' he observed in his journal, 'which others enjoy in other places, have Made these seeke to countervaile itt in home delights, as in ... little gardeins [and] Flower potts ... in which latter very curious rare rootes, plantts, Flowers, etts.'

Dutch villagers, too, enjoyed the delights of horticulture. At the height of the Golden Age, even the smallest settlements generally had flower growers' clubs, each with their own rules and fêtes. Most held a spring flower festival, where, just as today, different varieties would be placed in competition and prizes distributed. The festivals generally ended with a banquet held in honour of the winning flowers (any excuse for another feast, as foreign observers sourly remarked). Gardening had, in short, become something of a national passion.

Some time before 1635, the very first florists began to realize a profit from what had probably been rather tentative initial investments in flower bulbs. Word of their good fortune spread, and a few more newcomers decided to try their luck in the tulip trade as well. The writers and pamphleteers of the time are unanimous in stating that many of the incomers were weavers, who enjoyed certain advantages over other artisans in that their looms were worth a fair sum and could be pawned or mortgaged to raise the seed capital needed to enter the bulb trade, but they were soon joined by men from other occupations and even members of the professional classes, such as lawyers and clergymen.

Almost every sort of artisan had the desire to get rich, and at least some of them would have had the capital required to make a modest investment in bulbs. The chancers would have had less money, but greater willingness to risk what they did possess. Here two of the most striking characteristics of Dutch society came into play: the urge to save and the urge to gamble. These impulses may seem quite contradictory, but in fact they worked together to fuel the tulip mania.

Many visitors to the United Provinces were struck by the national horror of living beyond one's means, which, when combined with the general increase in wealth which the Republic enjoyed between 1600 and 1630, meant that (perhaps uniquely among all the people of Europe in this period) a significant number of Dutch families had savings. Because there were no banks, in the modern sense, in the Republic, we have no idea what sort of figures were typical, but Sir William Temple, for one, appears to have thought that a frugal Dutchman might save a fifth of his total income. If we take that as a guide, then a reasonably well-off artisan earning between 300 and 500 guilders a year might be expected to have had between 60 and 100 guilders a year to invest. Of course the working classes lived much closer to the poverty line than the merchants Temple had in mind when he made his estimate, so it is probably rather optimistic to use even his rough figure; but even so, it was surely possible that a family in which both parents worked consistently and tried hard to save money might scrape together 20 or 50 guilders at the end of a good year. In normal times that money would probably be spent on luxuries such as linen, household furniture and a few pieces of china, but even after tulip prices had appreciated throughout the 1620s it would also have sufficed to buy a few bulbs.

Like saving, the urge to gamble also infected all classes of society. No Dutchman, the businessman Willem Usselincx said, would put his money into an old sock when he could use it to make more money. For a rich merchant, this might have meant investing what he could afford in a risky voyage to the Indies. For the rest of society, betting was often a product of the difficulties, which have already been noted, that many Dutchmen experienced in trying to better themselves in an overcrowded country. Lotteries, for example, were as popular in the Holland of the Golden Age as they are today, and for many people winning some wager seemed an enticingly simple way of making some money.

The Dutch were notorious for their addiction to gambling. The French traveller Charles Ogier wrote that it was impossible to find a porter to carry one's luggage at Rotterdam, because as soon as a visitor had chosen one, another would arrive and play dice with the first for the client's business. Contemporary records mention that a man called Barent Bakker won a life-threatening bet that he could sail in a kneading trough down the Zuyder Zee from the island of Texel to Wieringen, and a Bleiswijck innkeeper named Abraham van der Steen lost his house on a wager concerning the precise appearance of a specific pillar in Rome. Dutch soldiers were even observed making odds on the outcome of battles which were still in progress.

Compared to such insane wagers, tulips looked a good investment. Growing bulbs was a lot easier than spending an eighty-hour week hammering horseshoes or working a loom, and because demand for the flowers was steadily increasing, prices consistently rose, at least for the finer varieties. No wonder Dutchmen thought they had chanced upon the dream of every gambler: a safe bet.

CHAPTER 9

Boom

Deep inside the long, low-lying cordon of islands that
separated the northern provinces of the Dutch Republic
from the North Sea stood the West Friesland town of
Hoorn. It was a port of moderate size, built upon a sheltered
bay which faced south on to the Zuyder Zee, the huge
inland sea which cut the United Provinces almost in two.
Until the 1550s, Hoorn had been one of the most important
places in the Netherlands, thriving on the Baltic trade. Now,
nearly one hundred years later, the ships that had once
unloaded cargoes of hemp and timber in its docks sailed on
to Amsterdam. Hoorn was dying; the port had slipped into
a long, slow decline from which it was never to recover.

Somewhere in the centre of this ruined town, in the first
half of the seventeenth century, stood a substantial house
with three stone tulips carved into its façade. There was
nothing else special about the building, apart from the fact
that it was eventually converted into a Catholic church. But
this was where the tulip mania began.

The stone flowers were placed there to commemorate

the sale of the house, in the summer of 1633, for three rare tulips. It was in this year, according to the chronicle of a local historian named Theodorus Velius, that the price of bulbs reached unprecedented heights in West Friesland. When news of the sale of the tulip house got out, a Frisian farmhouse and its adjoining land also changed hands for a parcel of bulbs.

These remarkable transactions, which took place in a part of the United Provinces that had been badly battered by recession, were the first sign that something untoward was beginning to happen. For three decades, flower lovers had used money to buy tulips. Now, for the first time, tulips were being used as money. And they were being valued at huge sums.

It is difficult to be certain how significant the sale of the tulip house was without knowing what sort of flowers were involved in the sale. But even though the price of homes in West Friesland might not have been high compared to those in Amsterdam, a decent-sized house within the walls of Hoorn would hardly have changed hands for less than 500 guilders or so, and good-quality farmland would probably have been more expensive than that; the value of each bulb would therefore have been high by the standards of the time. It is true that bulb prices had been rising for some years before 1633, and some equally startling transactions of which no record has survived may have taken place in earlier years; it is also likely that if a farm really did change hands for some bulbs, the man who sold it was a connoisseur landowner who possessed many other properties, and who passed this one on to an equally wealthy acquaintance complete with a sitting tenant, rather than a farmer disposing of his only means of earning a living. Yet, even so, these transactions were on a much larger scale than almost anything which had taken place in the 1620s.

The flower trade was changing, too. The bulbs that were bought and sold in the 1630s were not out-and-out rarities such as Semper Augustus, which could not be obtained for any sum, but other superbly fine varieties and, later, tulips of a lesser quality, most of which – though available only in limited numbers – could be bought from professional growers who would sell them to anyone who could pay their prices. As the number of people attracted to the bulb trade increased, the price of the most favoured varieties began to rise: slowly at first, but more rapidly from the end of 1634. This acceleration continued through 1635 until, by the winter of 1636, the value of some bulbs could double in little more than a week.

The tulip mania was to climax in just two mad months: December 1636 and January 1637. In those few weeks, people and money poured into the tulip trade as Dutchmen across the United Provinces rushed to invest whatever they possessed in bulbs. Naturally, this sharp increase in demand pushed prices higher still. For a while, at least, everyone made money. And that attracted yet more novice florists to the trade.

A contemporary chronicler gave some idea of the way in which prices rose within the space of two or three months: an Admirael de Man which had been bought for 15 guilders was resold for 175; one of the *Bizarden*, Root en Gheel van Leyde, increased in value twelvefold, from 45 guilders to a princely 550, and a Generalissimo tenfold, from 95 guilders to 900. The price of another superbly fine tulip, Generael der Generaelen van Gouda – the highly coveted 'General of Generals', a large flower with flaming scarlet stripes on a white ground whose unwieldy title was soon abbreviated, simply, to 'Gouda' – rose by two-thirds between December 1634 and December 1635, then by a further 50 per cent in the six months between December and May 1636. After that it tripled in value once again between June 1636 and January 1637, so that a bulb which was already expensive, priced at 100 guilders at the beginning of the boom, was worth no less than 750 just two years later.

Naturally, the prices quoted for a single bulb of the most celebrated of all tulips, Semper Augustus, had risen sharply too – from 5500 guilders a bulb in 1633 to an astonishing 10,000 guilders in the first month of 1637. The last sum mentioned could only have been afforded by a few dozen people in the whole of the Dutch Republic. It was enough to feed, clothe and house a whole Dutch family for half a lifetime, or sufficient to purchase one of the grandest homes on the most fashionable canal in Amsterdam for cash, complete with a coach house and an 80-foot garden – and this at a time when homes in that city were as expensive as property anywhere in the world.

Such profits were startling. Those who tried the bulb trade and profited from it could not resist telling their friends and family about the source of their good fortune; the novelty and the implausibility of making money from flowers ensured that their stories were told and retold – losing, it is certain, nothing in the process. By the end of 1634 or the beginning of 1635, lurid tales of the money to be made in tulips were the talk of Holland.

One such anecdote mentioned a piece of farmland on

Schermer polder which changed hands for half a dozen flowers; another told of a man who was so addicted to the tulip trade that the woman he had planned to marry left him for another. A third story concerned a rich merchant from Amsterdam who was said to have purchased a fabulously rare *Rosen* bulb, which he put down for a moment on a counter in his warehouse. When he looked again he discovered it had vanished, and his servants turned the place upside down in their search for the flower without success. Finally the merchant realized it must have been taken by a sailor – just returned from a three-year voyage to the East Indies and completely ignorant of the tulip craze – who had been in the warehouse at the time. He scoured Amsterdam for the man, and eventually found him sitting on a coil of rope down at the docks and chewing on the last portions of the precious bulb, which he had mistaken for an onion. When the merchant realized what had happened, he had the sailor seized and thrown into prison. A fourth tale was told of an English traveller, similarly ignorant of tulips, who used his pocket knife to dissect a bulb he found lying in the conservatory of his wealthy Dutch host. Unfortunately for him, it proved to be an Admirael van der Eijck (a *Rosen* variety adorned with exceptionally strong, straight blood-red stripes) worth no less than 4000 guilders. The inquisitive Englishman, too, soon found himself hauled before the magistrates and made to pay for his transgression. Or so the story went.

In truth, these and the welter of other anecdotes which circulated about the tulip trade were implausible at best, impossible at worst. Many were nothing more than common gossip and the rest appear to have begun life as simple morality tales, spun perhaps in pulpits, which warned of the dangers of dealing in flowers. But if they were intended to deter people from dabbling in tulips, such tales of excess were anything but effective. They made bulbs seem desirable, profit as certain as importing a cargo of nutmeg or a consignment of porcelain. Excited talk about the money that could be made in the tulip trade drove more and more people to try it for themselves.

What made so many people, from so many different occupations, so keen to try their luck in a trade of which almost all of them were completely ignorant? The lure of profit, certainly, and the prospect of making far more money than they ever had before. It helped, too, that the United Provinces were just emerging from a lengthy recession – which lasted for most of the 1620s and was the worst of the entire seventeenth century – caused in part by the renewal of the war with Spain and the effects of a Spanish blockade. This depression was followed by an increasingly feverish boom in the Dutch economy as a whole, which began in 1631 or 1632 and gathered pace towards the end of the decade and meant that in many cases there was more money around than ever before. Much more local factors, however, also had an impact. Many of the weavers who were drawn to the bulb trade came from the town of Haarlem, a dozen miles to the west of Amsterdam, where even the general boom could not prevent the linen business from falling into sharp decline as Leiden came to dominate the Dutch cloth industry.

Another influence was a severe outbreak of bubonic plague which struck many Dutch cities between 1633 and 1637. The chronicler Theodorus Schrevelius, who lived in

Haarlem throughout this period, recorded that the disease killed 8000 of his fellow citizens between its first appearance in October 1635 and its eventual disappearance in July 1637. Of these, more than 5700 died of plague between August and November 1636: one in eight of the total population of the city, so many that there were not graves enough to hold the dead. The appalling impact of the plague had two significant consequences. One was that it created a shortage of labour, and thus resulted in a rise in wages as employers competed for manpower; this would have helped to create surplus income which could be ploughed into the bulb trade. The other — or so it has been suggested — was to create a mood of fatalism and desperation among the traders themselves, which may have contributed to the abandon with which they dealt their bulbs.

Whether they were optimistic or fatalistic, the novice florists who did decide to try their luck in the tulip trade could hardly have hoped to possess a flower as valuable as a Gouda or an Admirael van der Eijck; they would have begun by buying and selling the cheapest available bulbs. The historian Simon Schama has suggested that newcomers were able to gain a foothold in what was already an expensive market because the professional growers happened to introduce an unusually large number of new varieties in 1634 and this had the effect of depressing prices. There does not appear to be any direct evidence that this was the case, and anyway it was the newest — and thus scarcest — varieties that were generally also the most expensive. What seems more likely is that some of the older and more established tulips had multiplied by this date to the point where they became generally available and modestly priced. It was by buying and selling these flowers that the newcomers must have entered the market.

Entering the tulip trade was simple. Investing in a few bulbs meant having a little money and access to a nearby nursery, but little else. In the first half of 1635, then, the market for bulbs began to flourish as never before throughout the United Provinces, springing up wherever tulips were readily available. Groups of florists emerged in every town where connoisseurs or growers were already well established: in Haarlem and Amsterdam; in Gouda and Rotterdam; in Utrecht and Delft, Leiden and Alkmaar; and in Enkhuizen, Medemblik and Hoorn.

The growers and the connoisseurs did more than simply provide the newcomers with stock. The trade they had created was already ordered and established. There were no arcane laws to master, no complications to be overcome. The rules for buying and selling flowers were based on simple common sense, and they were well known and accepted long before the first florists began dealing in tulips.

The earliest sales were probably by the bulb, but this changed as the number of flowers available increased, and it would appear that by 1610 some less valuable tulips were already being sold 'by the bed', a unit of exchange which does not seem to have been precisely defined. The legal archives of Haarlem contain the record of the sale, in 1611, of four beds of tulips planted by an apothecary called Joos to one Jan Brants, who paid the already generous sum of 200 guilders. The next year Brants bought two more beds of tulips which belonged jointly to a certain Dammis Pietersz. and a Haarlem brewer named Augustijn Steyn. They cost him another 450 guilders.

Some time after that (when is not clear), it became possible to buy and sell offsets as well as mother bulbs. This was an obvious next step, because logic dictated that offsets, which after all would soon become bulbs themselves, must be worth something in their own right. Nevertheless, this extension to the trade was fraught with difficulty because it was impossible to guarantee that offsets would mature satisfactorily or, as we have seen, that the tulips they produced would be identical to those of the mother bulb. Because of these problems, trading in offsets was something of a risk, and the idea took some time to win favour. When, in the spring of 1611, a Haarlem connoisseur called Andries Mahieu was asked if he would sell a linen merchant of his acquaintance some offsets, he replied by asking his friend if he really wanted to buy 'a cat in a bag'. This statement so imprinted itself in the mind of one bystander, the gardener Marten de Fort, that it survived to be recorded in the legal archives too.

Trading offsets was significant for another reason. Clusius and the other early growers already knew that bulbous plants prosper best if they are lifted from the soil soon after the flowers of one season have fallen, then dried off and stored above ground until autumn – usually on shelves which allowed the circulation of air. The buying and selling of bulbs therefore occurred only during the summer months when the tulips were out of the ground and could be physically exchanged. Offsets, on the other hand, take several years to mature, so it was tempting to sell them when they first appeared.

Dealing in offsets was the first step to liberating the tulip trade from its traditional dependence on the calendar. It meant that some of the buying and selling which had previously been crammed into no more than four months could now be spread throughout the year. By itself, the sale of the odd offset a few months before it was actually ready to be separated from its mother bulb posed no threat to the stability of the tulip trade. But it set a dangerous precedent, and as more and more florists came flooding into the market, the pressure to make tulip dealing a year-round affair only grew.

A trading season which ran only from June until September made perfect sense to the connoisseurs, who preferred to see a plant in flower before they considered buying it and wanted to complete all their purchases for the year in time for the bulbs to be returned to the flower bed. But it was highly limiting for the new breed of tulip dealers. Because they generally had no interest in cultivating their bulbs, the old distinctions between the growing season and the lifting season meant little to the florists, who took less pleasure than their predecessors in the physical beauty of the tulip and more in its potential to earn them money. The newcomers wanted to wring as much profit from their flowers as they could, and while a handful may have appreciated the benefits of planting the bulbs and making money from their offsets, most were far more interested in buying tulips simply to sell them on.

From the autumn of 1635, then, the bulb trade changed fundamentally and for ever. Ignoring the customs of the connoisseurs, increasing numbers of florists progressed from trading only tulips which they had in their possession to buying and selling flowers which were still in the ground.

Bulbs then ceased to be the unit of exchange; now, the only thing which changed hands was a promissory note – a scrap of paper giving details of the flower being sold and noting the date on which the bulb would be lifted and available for collection. To avoid chaos, a sign planted in the ground above each individual bulb recorded its variety, its weight and its owner.

There were advantages to the new system. It certainly permitted trading to take place throughout the months of autumn, winter and spring; and because the bulbs stayed where they were until lifting time no matter who their new owner was, it was very appealing to florists who had neither the skill nor the desire to cultivate bulbs themselves. But it was potentially very dangerous too. Buyers had no opportunity to inspect the bulbs they were buying or of seeing them in flower. There was no guarantee of quality. And a florist could not be sure that the bulbs he was purchasing really belonged to the seller, or even that they actually existed.

The Dutch called this phase of the tulip craze the *windhandel,* which can be translated as 'trading in the wind'. It was a phrase rich in meaning. To a seaman, it meant the difficulties of navigating a ship steering close to the breeze. To a stockbroker it was a reminder that both the tulip traders' stock and their profits were so much paper in the wind. To the florists, however, the *windhandel* meant trading pure and simple, unregulated and unconfined.

It was this innovation that made the greatest excesses of the mania possible. The introduction of promissory notes did much more than make the tulip trade a business which could flourish all the year round; it turned dealing into an exercise in speculation, and – because delivery was usually months away – it encouraged the sale and resale not so much of bulbs as of the notes themselves.

Exquisite flowers now became nothing but abstractions for dealers who cared only for their profits, and the repeated transfer of a dubious claim to ownership from one dealer to another became the chief characteristic of the bulb trade. Before long, to the scandal of strait-laced contemporaries, it became perfectly normal for florists to sell tulips they could not deliver to buyers who did not have the cash to pay for them, and who had no desire to ever plant them.

By agreeing to purchase bulbs that would not be ready for delivery for several months, the tulip traders had created what would today be called a futures market – simply defined, a form of speculation in which a dealer gambles on the future price of some commodity, whether it be flower bulbs or oil, by promising to pay a specified price for the goods on a fixed date some time in the future. This was an event of some historical significance. In the 1630s, the whole concept of futures was still a novelty. The very first futures markets had been organized in Amsterdam less than thirty years earlier, the invention of merchants who traded in timber, hemp or spices on the Dutch stock exchange. Tulips were the first commodity to be bought and sold outside the markets of Amsterdam, and the first to be traded by anyone other than high-ranking merchants and stock-exchange specialists.

This, of course, was a large part of their appeal. By 1635, the regents and the great merchants of the United Provinces could choose to invest their money in a variety of ways. They could earn guaranteed interest by buying government bonds or depositing their cash with one of the many new banks which were springing up. If they felt a little more adventurous, they could buy shares at the stock exchange or purchase a stake in a local drainage project or in a ship off to trade in the Americas. Each of these investments, though, required a substantial amount of capital, and so far as the artisans, the tradesmen and the tenant farmers of the Republic were concerned, it was all but impossible to find a profitable way of investing the little money that they had. There were no building societies in the seventeenth century, no unit trusts, no personal equity plans, no penny shares, no tax breaks and no tax shelters. For a Haarlem weaver, investment meant buying more flax or making a down payment on a new loom. Now, suddenly, there was a new way of making money – one that seemed alluringly straightforward, appeared to guarantee a profit and, above all, required little in the way of capital.

Futures trading is a highly speculative way of doing business, but it has significant advantages. It satisfies a seller, who might, for example, be awaiting a cargo due to arrive from overseas, and who is at any rate himself probably not yet in possession of whatever he is selling; he in effect sells the risk that the price of his goods will fall before he can get them to market. He can demand a deposit, say 10 per cent, of the agreed price, and being guaranteed a definite sum of money on a fixed date, he can arrange his finances accordingly. It can also be a highly profitable arrangement for a buyer, so long as he guesses correctly whether prices will rise or fall. For example, a florist who offered 100 guilders for a promissory note guaranteeing him ownership of a Gouda when it was lifted in four months' time wagered that he would be able to sell the note for more than that amount before he became liable to pay for the bulb. If he could actually get no more than, say, 80 guilders for his piece of paper, he would of course lose 20 guilders come lifting time, but in the constantly rising market for tulips, gambling on future prices must have seemed absurdly simple, and the chances of actually making a loss would have struck most of those who now flocked to buy bulbs as remote.

In truth, though, futures trading was anything but simple, and very much riskier than it at first appeared. Indeed it was exceptionally dangerous. A florist with a capital of only 50 guilders who was certain prices would continue to rise might, for instance, throw caution to the wind and agree to purchase five of the 100-guilder Goudas. His money would be enough to pay a deposit of 10 per cent on each bulb, and if by lifting time the price of the tulips had doubled, his 50 guilders would have made him the owner of 1000 guilders-worth of bulbs. After selling the flowers at the new and higher price, he could pay the balance of his obligation and walk away with a clear profit of 500 guilders. Thus, if the trade remained buoyant, poor artisans could indeed hope to make huge fortunes from flower bulbs. But should the price of tulips fall, catastrophe was certain and bankruptcy all but inevitable. If Goudas halved in value, for example, the florist who had invested his entire savings of 50 guilders in bulbs would be facing a loss of 200 guilders – a sum he could not possibly hope to pay.

The Dutch government had long been acutely aware of the risks of 'selling short', as it is known. Indeed, it had consistently ruled that trading commodities which were not in the possession of either the buyer or the seller was not merely dangerous but fundamentally immoral. Less than two years after the practice was introduced in 1608, it was banned, and laws repeating the prohibition on futures trading were passed in 1621, 1623, 1624, 1630 and 1636. The trade in tulip futures which developed in the 1630s was thus technically illegal, but the fact that the parliament of the United Provinces made six separate attempts to stamp the practice out amply demonstrates how little chance any such ban had of being properly enforced.

Selling short, then, was dangerous, even when the goods concerned were as uncomplicated as a cargo of Baltic timber. But tulips were an unusually volatile commodity, even by the elastic standards of the futures trade. A merchant who dealt in timber knew precisely what he was buying. A florist purchasing a tulip for delivery at lifting time had no idea. He was gambling on a living thing. To be successful, he needed not just a shrewd understanding of the price his bulb might command in several months' time, but some idea of what was happening to it while it was still in the ground.

The best way of making money on a flower was to buy one that was about to develop offsets which could be removed and sold separately. Bulbs that were likely to grow rapidly were thus more valuable than either immature flowers or those which were already fully developed and unlikely to produce more than a few more offsets before they died. But even the most experienced growers found it difficult to predict accurately what a single bulb of one particular variety would do, and so far as novice florists were concerned, bulb dealing was an exercise in pure speculation.

In order to give tulip traders the basic information they

needed to guess how a bulb might develop after planting, it became customary to indicate the weight of each bulb when it had been returned to the ground. Weights were given in *azen* ('aces'), an extremely tiny unit of measurement borrowed from the goldsmiths. One ace was equal to rather less than two-thousandths of an ounce – one-twentieth of a gramme – and mature tulip bulbs might weigh anything from 50 aces to more than 1000, depending on the variety. As well as indicating the date on which a flower would be ready for lifting, then, the promissory notes which were exchanged by florists also noted the bulb's weight when planted, and the ledgers each dealer used to record his purchases always included a column in which the dealer listed the size of his bulbs in aces.

From this it was only a short step to selling tulips not by the bulb but by the ace. In one respect, this had the desired effect of making trading fairer. Under the old system of paying by the bulb, a florist would have been charged the same for an immature tulip weighing, say, 100 aces, which might not produce offsets for another year or more, as he would for a mature specimen of 400 aces. Paying by the ace, he was charged a price that more accurately reflected the development of the bulb. But the new system also meant that prices increased much more rapidly than before. Most tulips increased substantially in size while they were in the ground, so even if the price charged per ace for a given variety did remain completely unchanged from the moment a flower was planted in September or October until it was lifted the following June, the value of the bulb was still almost certain to increase significantly.

The records of the tulip trade offer examples of just how

dramatically the money invested in a single bulb could multiply. A Viceroy grown by an Alkmaar wine merchant named Gerrit Bosch in his garden just outside the city walls weighed 81 aces when it was planted in the autumn of 1636. It had grown to 416 aces when it was lifted in July 1637 – a fivefold increase. An Admirael Liefkens in the same garden grew from 48 aces to 224 and a Paragon Liefkens from 131 aces to 434. Had the prices paid per ace for these three varieties remained unchanged, Bosch's customers would have enjoyed returns varying from 330 per cent to as much as 514 per cent in a scant nine months. There was probably not another investment in the whole of the United Provinces that offered such spectacular results this quickly, and certainly none that all but guaranteed it. A single round trip to the Indies took two years or so to complete, and while they were away the Dutch East India Company's ships were exposed to the dangers of disease, shipwreck, piracy and Spanish attack. Even the rich trades, then, exposed the privileged few permitted to invest in them to risks unknown to Holland's florists.

The earliest record of selling by the ace dates to the beginning of December 1634, when the Haarlem grower David de Mildt went with a linen worker named Jan Ocksz. to the garden owned by Jan van Damme on the Kleine Houtweg. On de Mildt's advice, Ocksz. purchased two red and white Goudas weighing 30 aces for 30 stuivers – one and a half guilders – per ace. He also bought two Admirael van der Eijcks, paying not by the ace but 132 guilders for each tulip, which suggests that the old system of dealing by the bulb was still in use in 1634. By 1635, however, all surviving records refer to bulbs sold by the ace.

As the tulip trade grew in confidence and complexity, florists occasionally elaborated on this basic system. It was, for example, possible to purchase bulbs on condition that they had reached a minimum weight by the time they were lifted. In another case which involved David de Mildt, a Haarlem clog-maker named Henrick Lucasz. bought two tulips – a *Rosen* variety, Saeyblom van Coningh, and a *Violetten* called Latour – at an auction organized by one Joost van Haverbeeck at the end of October 1635. With de Mildt as a witness, Lucasz. agreed to pay 30 guilders for the Saeyblom and 27 guilders for the Latour, but with the guarantee that the bulbs would weigh at least 7.5 aces and 16 aces respectively when lifted. In the event, they were found to weigh no more than 2 aces and about 13 aces, so Lucasz. asked van Haverbeeck to return the money he had paid in advance. Van Haverbeeck, a Haarlem dealer possessed of a notoriously short temper, indignantly refused to make a refund and the matter ended up in the hands of a solicitor. (If anything, Lucasz. escaped relatively lightly in this case. The records of the time show that van Haverbeeck and his equally abrasive father repeatedly issued violent threats against some of their customers and were the principal suspects when the valuable tulips growing in de Mildt's garden were vandalized in the winter of 1635.)

Other variations were also possible. A few poorer florists bought shares in expensive bulbs. On one occasion, an Amsterdam grower, Jan Admirael, sold a half share in three bulbs to a customer named Simon van Poelenburch. On another, Admirael entered into a complicated deal with a dealer called Marten Creitser, agreeing to swap several tulips and 180 guilders in cash for eleven paintings and an engraving owned by Creitser.

Nevertheless, the introduction of pricing by the ace did not mean that tulips of any given variety cost the same everywhere within the Dutch Republic. Since even the most important messages could travel no faster than a man on horseback, there was no way to communicate changes in price quickly and accurately from place to place and thus no single market for tulips. Instead, each town involved in the bulb trade valued flowers slightly differently; some places were generally expensive, others cheap.

Other factors added to the general chaos in pricing. Not only did individual florists have preferences of their own; they were also influenced by which tulips had just been bought, which sold, and by which flowers were in fashion and which were becoming more easily available. Large bulbs were generally cheaper per ace than small ones – and when all of these factors were taken into account, even tulips bought in a single place on a single day could vary significantly in price. Seven Goudas sold in Alkmaar within the space of an hour or two fetched prices varying from 6 guilders, 3 stuivers per ace to 10 guilders, 2 stuivers, which meant the buyers paid from 765 guilders to 1500 guilders a bulb. Three tulips of a variety called Paragon van Delft were purchased within minutes for 1 guilder, 14 stuivers an ace, 2 guilders, 4 stuivers, and 4 guilders, 2 stuivers respectively, and bulbs of Admirael van der Eijck weighing 92, 214 and 446 aces sold for 710, 1045 and 1620 guilders apiece.

The rapid increase in bulb prices in 1635 and the first half of 1636 had important consequences. Growers and dealers who had hitherto traded bulbs only to connoisseurs or among themselves recognized that there were new opportunities to make money. They began to offer their flowers to the florists who were streaming into the market. Then, as a next step, some banded together so as to maximize their capital or improve the variety of stock they had to offer. A number of companies were formed to trade in bulbs. In September 1635, for example, the Haarlem merchant Cornelis Bol the Younger went into partnership with a grower called Jan Coopall, Bol contributing 8746 guilders, 2 stuivers to the capital of the company. And in December 1636, the Haarlemmers Henrick Jacobsz. and Roeland Verroustraeten went into business with Philips Jansz. and Matthijs Bloem of Amsterdam. The articles of their company explained in some detail how the business would operate. The 35-year-old Verroustraeten, who was probably already an experienced trader, was the only one authorized to deal in bulbs, and he would buy and sell tulips with money put up by the other three directors. All four directors agreed to trade only on behalf of the company, and never on their own account.

By the autumn of 1636, both the tulip companies and the professional growers must have been thinking carefully about what stock to plant for the next season. The most valuable flowers – the Admiraels, Generaels, Generalissimos and their kin – were already too expensive for many florists to afford, and the poorer traders at the bottom of the market had begun to ask for less favoured tulips which were available in greater quantity and were significantly cheaper. Like the superbly fine varieties which had been the basis of the bulb trade in the early 1630s, these flowers were termed 'piece-goods' – that is, tulips that were bought and sold as single bulbs – but because their prices were low they were quoted not by the ace but in multiples of 1000 aces. Varieties sold in this way included several that became famous later on, such as the vermilion-streaked Rotgans and Oudenaers and the unusual white-on-purple Lack van Rhijn. With their uncompromising, broader flares of colour, they were less cherished by contemporary connoisseurs and growers, but they would have seemed more familiar to modern gardeners than the rarer, lightly streaked fine tulips.

Some ambitious artisans had begun to buy and sell tulip bulbs in 1634 or 1635, but the legal records of the tulip trade suggest that as late as the summer of 1636 the majority of tulips were still being sold by their growers direct to customers who planned to plant them in their gardens. By the autumn, however, the market had been all but taken over by florists who bought and sold simply to make a profit, with the greatest influx of newcomers coming in December 1636 and January 1637.

They came from all walks of life. According to one contemporary pamphleteer (who may have been employing a certain amount of poetic licence), their numbers included bricklayers and carpenters, woodcutters and plumbers, glass-blowers, farmers and tradesmen, pedlars and charcuterers, confectioners, smiths, cobblers, coffee-grinders, guards and vintners – not to mention dry shavers, furriers and tanners, coppersmiths and schoolmasters, millers and also demolition men.

Few details have survived of the trading which took place as the boom in tulips became frenzied in the last two or three months of 1636, but a short series of pamphlets containing a fictionalized account of the tavern trade are agreed to be both reliable and representative of what actually occurred. These are the three *Samenspraecken tusschen Waermondt ende Gaergoedt* ('Conversations between Truemouth and Greedygoods'), written by an unknown author and published at the beginning of 1637 by Adriaen Roman, the principal printer then living in Haarlem.

The Gaergoedt of the pamphlets is a weaver who has abandoned his craft to become a florist. He has mortgaged all the tools of his trade to provide himself with working capital, and now travels from town to town dealing in bulbs. On a rare visit home, he meets his old colleague Waermondt, who has yet to become involved in the burgeoning craze, and offers him wine and beer. Then Gaergoedt attempts to persuade his friend to enrich himself by buying and selling tulips. At present, he points out, Waermondt struggles to make a profit of 10 per cent on his business. With tulips, he will make 100 per cent or more: 'Yes, ten for one, a hundred for one, and sometimes a thousand.'

The *Samenspraecken* take a predictably moralistic view of the tulip trade. Gaergoedt is hubristic and sublimely, stupidly confident that the price of bulbs will go on rising for ever. He boasts that he has already earned a fortune from his flowers and pays his way through life with bulbs. His friends – other weavers and gardeners – are also rich and drive from town to town and from college to college in richly decorated coaches.

Waermondt, whom the anonymous pamphleteer casts in the role of bemused but honest beginner, finds it hard to believe that a mere weaver can make such sums, and under his questioning Gaergoedt is forced to admit that he has yet to receive most of the money due to him as a result of his successful trading. His profits cannot be realized until the tulips are lifted again the next summer. Still, he says, 'this trade goes steady', and another two or three years in the bulb market will more than set him up for the rest of his life. Then, he says, he will use his profits to buy a brewery, a bailiwick, even a lordship.

Waermondt is incredulous; the whole thing, he thinks, is just too good to be true. He wonders how the common people caught up in the tulip craze dare risk all the money they are borrowing against the profits of the trade. And, though he is certainly tempted by the talk of money, he tells his friend he prefers not to take the risk of plunging into the flower business.

In the autumn of 1636, many Dutchmen must have thought, like Waermondt, that the profits being made on tulips were simply too good to be true. But thousands did not, and they took their savings and mortgaged their goods in order to take part in the hurly-burly of the bulb trade.

Most had little access to ready money, but the traders and florists who were already in the market saw an opportunity to sell their flowers to novices who had little understanding of which tulips were valuable and which were not, and it quickly became customary to accept deposits not in cash but in kind. For florists whose wealth – what there was of it – was tied up in their possessions, this meant paying for bulbs with whatever came to hand. The fictional Gaergoedt offered deposits ranging from cloth enough to make a coat and suit to a quarter of prunes. Real florists paid in tools, clothes and household goods if they were artisans, farm animals or crops if they were farmers, paintings and other luxuries if they were rich. The balance of the purchase price was payable only on delivery, which took place at lifting time. On occasion payment terms could be even more flexible: one agreement, in which the Haarlem shopkeeper Aert Ducens sold his entire garden to a local gentleman named Severijn van de Heuvel for 1600 guilders, specified that payment would fall due only on New Year's Day 1638, a full year after the contract was agreed.

The *Samenspraecken* give further examples of the sort of agreements struck by these inexperienced tulip traders once the idea of paying deposits in kind became generally accepted. As Gaergoedt talks his friend Waermondt through the deals he has made and noted in his ledger, he points out one in which he sold a packet of a variety called Witte Croon ('White Crown') for 525 guilders in cash, with a deposit of four cows to be paid immediately, and another in which he purchased a quantity of Genten bulbs by handing over a deposit of 'my best shot coat, one old rose-noble, and one coin with a silver chain to hang around a child's neck' and agreeing to pay 1800 guilders cash when the bulbs were ready for delivery. Some agreements appear to have been even more complicated than that. For example, the *Samenspraecken* suggests that florists sometimes offered bulbs of one variety in part exchange for tulips of another. One of Gaergoedt's most extravagant arrangements called for him to receive a large quantity of Witte Croonen, together with a coach and horses, two silver bowls and 150 guilders cash. On his part, the weaver agreed to hand over a silver dish worth 60 guilders, an equal amount of Gheele Croonen ('Yellow Crowns'), and 200 guilders in cash.

As the autumn of 1636 shaded into winter, all seemed

well in the flower business. The number of florists and the number of bulbs in circulation both continued to increase. Prices rose steadily. Profits were enormous. Yet in reality, the tulip trade that the florists had built rested on the shakiest of foundations.

It was not simply a matter of whether the market could possibly sustain the rapid rise in bulb prices. All sorts of problems occurred when a florist was unable to inspect the flowers he was purchasing. To begin with, there was no guarantee that the tulips were being handled with proper care. The Haarlem archives contain the details of a case concerning a local baker named Jeuriaen Jansz., who in the spring of 1636 found a beautiful specimen of Admirael Liefkens flowering in the Amsterdam garden of Marten Creitser. Jansz. struck a deal to buy the offsets. A few months later, the baker was sitting in a tavern college when another florist told him the bulb had been lifted prematurely, and thus might have been damaged. Jansz. had to threaten legal action to force Creitser to release him from his obligation to purchase the offsets.

Even rich connoisseurs ran the risk of buying damaged goods. Cornelis Guldewagen, who had been one of the aldermen of Haarlem, acquired no fewer than 1300 tulips from Anthony van Flory of The Hague, and retained Barent Cardoes to plant them in his garden outside the Cruyspoort by the city moat. When the bulbs were unpacked, Cardoes and his assistant found they had been lifted very clumsily and about half had been badly damaged.

The poorly understood mysteries of breaking also caused considerable problems. Anyone who purchased an offset risked buying a breeder bulb rather than the broken tulip he desired. In May 1633, Abraham de Goyer, one of the most prominent tulip dealers of Amsterdam, bought two Paragon Schilders at an auction organized by the man who had created the variety, Abraham de Schilder himself. Paragon Schilder was a new variety, and highly coveted; judging by the date that de Schilder chose to hold his auction, de Goyer had probably seen the tulip in flower a few days earlier and been entranced by it. At any rate, he paid what was by the standards of the time a substantial price for his two bulbs – 50 guilders for one and 41 guilders for the other – planted them in his garden just outside the city walls, and settled back to wait nine long months for them to bloom again. Finally, in the spring of 1634, the longed-for tulips flowered – but when they did, the two Paragons proved to be nothing like the glorious *Rosens* that de Goyer had anticipated. The pure whites and vivid scarlets that the grower had fallen in love with in de Schilder's garden were nowhere to be seen; De Goyer's 90 guilders had bought nothing but the muddy colours of inferior breeders. The unfortunate grower was still demanding his money back eighteen months later, even though it was generally accepted that reputable bulb dealers would consider a purchase null and void when an offset failed to match the quality of the mother bulb.

Most serious of all were a handful of cases of outright fraud, which were perhaps inevitable in a market as rich and poorly regulated as the bulb trade. When tulips of the same variety could often differ quite substantially in appearance, and a poor Viceroy could look much the same as a less valuable *Violetten*, say an Admirael van Engeland, it was often difficult to distinguish between real deceit and genuine mistakes. Certainly the legal archives of the Dutch Republic appear to contain few cases which were proven. But Waermondt, in the *Samenspraecken*, says he has spoken to his cousin, who has experience of the tulip trade, and been told of people who paid for Witte Croonen and received instead worthless common tulips. Of course, because all bulbs looked much the same, frauds such as this were discovered only when the tulips flowered in the spring.

But though problems like these concerned more conservative and cautious Dutchmen, the florists who flocked to trade in tulips in the autumn of 1636 focused almost solely on the money they were making. Because demand for bulbs was growing day by day, prices were rising more and more quickly; by this time, the contemporary chronicler Lieuwe van Aitzema recorded, everything that could be called a tulip – even bulbs that had been considered so useless they had been thrown away on dunghills only months before – was now worth money.

In most respects, all that was required for the boom in tulip prices to turn into a full-fledged mania was now in place. Many different varieties had been created, some much coveted but scarce, others less desirable but easier to obtain. A small group of professional gardeners existed to breed new flowers and supply at least some of the demand for the existing ones. A larger group of competent and enthusiastic amateurs, certainly several hundred strong, were also growing tulips in their own gardens, so the flowers could already be found in almost every town. The rules of trading had been established, and there were criteria for measuring a flower's worth and allotting it a place in a scale which ran from superbly fine to rude. The traders and growers who dominated the trade had been joined by thousands of florists willing to sell everything they owned for bulbs. Finally, prices were higher than they had ever been before. All that was needed now was a way of bringing aspiring tulip dealers together: a place in which to trade.

At the Sign of the Golden Grape

CHAPTER 10

At the Sign of the Golden Grape

Right in the heart of Amsterdam, almost on top of the dam that actually gave the town its name, was an elegant four-storey quadrangle, built in the Flemish style and crowned with a slim clock tower. This building stood opposite the central bank and close to the town hall in a position which emphasized the central role it played in the life of the city and indeed the United Provinces as a whole. It was Amsterdam's new *beurs* – the city's stock exchange.

Not too many years before, the traders who now occupied one or other of the 123 offices in the exchange had been forced to transact their business out in the open on Amsterdam's New Bridge or – if wet – among the pews of St Olaf's Chapel or the town's Old Church. As the city boomed in the early years of the seventeenth century, however, and foreign trade poured in, it became clear that the stock exchange needed a permanent and weatherproof home. The *beurs*, which opened for business in 1610, met that need and, by its sheer physical presence, went some way to assuaging the suspicions of Amsterdam's more conservative burghers, who felt there was something faintly ungodly about dealing in shares.

Trading on the *beurs* was strictly regulated, and was permitted only between the hours of noon and two. Each day's business had to be packed into those two hours, and the raucous frenzy that erupted within the quadrangle as the big clock in the tower struck for midday was such that anyone strolling past the exchange at noon might be forgiven for concluding that the burghers had a point. Trading was conducted at such a pace that brokers who years earlier had sealed each deal with an elaborate ritual of handshakes now merely slapped wildly at each other's hands before rushing on to the next.

Hundreds of traders were licensed to deal on the stock exchange. There were perhaps four hundred official *beurs* brokers in the 1630s, and they were joined on the trading floor by up to eight hundred unlicensed freelance dealers who specialized in trading small packages of shares at low prices. In his description of the exchange, the contemporary writer Joseph de la Vega observed one such freelance dealer, who 'chews his nails, pulls his fingers, closes his eyes, takes four paces, and four times talks to himself, raises his hand to his cheek as if he has a tooth-ache, and all this accompanied by a mysterious coughing'. Vega does not mention what his small-time broker was hoping to buy or sell for his handful of guilders, but he had a considerable choice: by 1636, at least 360 different commodities were traded on the Amsterdam exchange, from precious metals to French brandy. Tulips, however, were not among them.

This fact may come as a surprise to those who assume that a financial calamity with the reputation that the tulip mania enjoys must necessarily have been not merely serious and widespread, but also have had a significant impact on the contemporary stock market, on trade, and on the Dutch economy in general. Nothing could be further from the truth. The speculation in tulip bulbs always existed at the margins of Dutch economic life. It was conducted by amateurs, not professional traders, and was never subject to either the customs (however peculiar) or the regulation of the stock exchange. In fact the mania took the form of a rough but intended parody of the trade in commodities and shares that flourished on the *beurs*. It was the province not of financiers experienced in the ways of business, but of country people and poor city dwellers who had, when they started dealing in bulbs, almost certainly never owned a single share in their whole lives.

The fact that the tulips were not dealt on the stock exchange does not mean the flower business was not regulated. In fact, it soon evolved into a complicated, even ritualized, affair in which buyer and seller dealt according to fixed rules and were united by mutual obligations, agreed in front of witnesses and noted down in writing. Like the brokers who once congregated on the New Bridge, the tulip traders needed somewhere to transact their business. Like the brokers, some of them used the house of God upon occasion; when the mania took place, the local church was a general meeting place pressed into use by everyone from local merchants to courting couples. Most, however, found it far more comfortable to buy and sell their bulbs in a convenient tavern. The tulip trader's stock exchange was his local pub, and unless the conditions in which the bulbs were actually traded are understood – late at night, in smoke-filled rooms, by drunken men – the mania itself will always remain a mystery.

Inns were, to begin with, so common in the United Provinces as to be commonplace. In 1613, Amsterdam already had five inns for every hundred inhabitants, which suggests that in 1636 there were probably two hundred packed within the city walls of Haarlem – an area not that much bigger than Hyde Park. These drinking houses ranged from full-fledged taverns to dingy cellars and apothecary shops. Perhaps a fifth were unlicensed and illegal, and specialized in evading the high beer tax imposed to help pay for the war with Spain. The authorities had to carry out frequent raids to keep the spread of such establishments in check.

It was only the larger and more reputable inns, however, that would have been able to offer the private rooms required by the tulip traders. They went by names such as the Beelzebub, the Finch, the Lion, and the Devil on a Chain. Establishments of this sort could be found both within and without a city's walls.

In Haarlem, many taverns clustered to the south of the city, amid the glades and walks of the woods. Because they were close to the earliest tulip farms just outside the walls, it seems reasonable to assume that some of them, at least, must have hosted groups of florists trading bulbs. If so, then the tulip dealers would have shared the premises with unsavoury companions. Prostitution having been outlawed – ostensibly at least – within Haarlem's city walls, the taverns of the Haarlemmerhout frequently doubled as brothels. The most notorious of the local whorehouses cannot have been easy to miss – it appears in the records of the time as 'the red house outside the gate of the cross'*.

We do not know for certain how many of the dozens of taverns in Haarlem itself played host to the tulip maniacs of 1636, but it seems a fair guess that one of them was a large and well-known inn called De Gulde Druyf ('The Golden Grape'), which occupied a prime location on the corner of the market square and the city's main street, the Koningsstraat. This tavern was owned by the brothers Jan and Cornelis Quaeckel, though they did not run it day to day. The Quaeckels were the sons of an innkeeper called Cornelis Gerritsz. Quaeckel, who had been one of the most important pioneer tulip growers in Holland. At least five new varieties of tulip, created by him in the first quarter of the seventeenth century, bore the Quaeckel name in honour of his achievements, including the white-and-violet Lack van Quaeckel and a popular yellow-and-rust *Bizarden* named Mervelye van Quaeckel – 'Quaeckel's Miracle'. Old Quaeckel died, aged almost seventy, in 1632, but his youngest son Jan continued to be active in the tulip business up to and beyond the peak of the mania. Nothing could have been more natural than for him to have played host to Haarlem traders in a back room of his own tavern, which was not only perfectly situated but also one of the most popular watering holes in Haarlem.

Suppose, then, that we were to travel from Amsterdam to pay a visit to the Golden Grape one day in the late autumn of 1636 and watch tulip traders at work. What would we see?

Leaving Amsterdam late in the afternoon and travelling,

* Or, in Dutch, the Cruyspoort.

perhaps, along the newly opened passenger canal which linked the two cities – the first of its kind in the United Provinces – visitors would arrive at Haarlem at dusk. The journey from one city to the other took only two and a quarter hours. It was so quick and so convenient that fashionable Amsterdammers found it easier to send their dirty washing by boat to the superior laundries of Haarlem than to do it themselves. Those on board the canal boats passed the time discussing current affairs and reading specially produced small pamphlets called *schuitepraatjes*, or 'towboat talks'. During the autumn and winter of 1636, the boats would certainly have been hotbeds of gossip about the latest developments in the tulip mania. As they approached, the visitors' first glimpse of Haarlem would be of a long line of red-brown roofs, crowned with wisps of smoke from many thousands of chimneys, rising clear of the meadowlands that surrounded the town. Next they would see that a low perimeter wall of brick and a defensive moat spanned by nine bridges protected the city. Far to the west, beyond the roofscape, the ragged outlines of the giant sand dunes which lined the North Sea coast might just be seen rising to meet the characteristic soft grey sky of Holland. And to the south they would glimpse the grim, black expanse of the Haarlemmermeer – huge, brackish and shallow, a windswept inland sea prone to violent storms, constantly eroding its banks and eating up more and more of the surrounding farmland so that now it stopped only a mile or so short of the walls of Haarlem itself. The mere enjoyed an evil reputation for claiming the lives of those foolish enough to sail on it; Haarlemmers called it 'the water wolf'.

Alighting from their barge just outside the city walls, travellers from Amsterdam would find themselves standing at a gate called the Amsterdamse Poort. Here Haarlem's regents had erected a set of gallows, a triangle of three brick pillars joined by iron beams, and some wooden posts to which were strapped the bodies of recently executed criminals. Because the city was the home of the official executioner for the whole province – a man who bore the title 'Master of High Works of Holland' and saw to the dispatch of prisoners from Amsterdam as well as Haarlem's own criminals – these contraptions would probably be full. When Sir William Brereton passed this way in 1634, he encountered not only the fleshless skeletons of two unfortunates swinging from the gibbet but also the mutilated body of a girl who had been broken on the wheel for murdering her own child, and the blackened corpse of a beggar who had been burned at the stake for setting a whole village ablaze.

Entering the city through the Amsterdamse Poort, the visitor would first notice Haarlem's distinctive smell. The city stank of buttermilk and malt, the aromas of its two principal industries: bleaching and beer. Haarlem breweries produced a fifth of all the beer made in Holland, and the town's celebrated linen bleacheries, just outside the walls, used hundreds of gallons of buttermilk a day to make cloth shipped to the city from all over Europe a dazzling white. The buttermilk filled a series of huge bleaching pits along the west walls, and each evening it was drained off into Haarlem's moat, and thence into the River Spaarne, dying the waters white.

Night draws in quickly in the Dutch Republic by late autumn, and the travellers from outside the city would have had to find their way to the market square in the dark. The only light in Haarlem's maze of cramped streets – some so narrow that the occupants of a house on one side of the road could reach across and shake hands with their neighbours on the other – came from fires and oil lamps gleaming through shutters barred against the cold. The town, so crowded and alive with noise during the day, would have been much quieter by night. With the exception of the ritualistic clatter of a militia company on guard, most of the roads would be deserted but for the hunched figures of drinkers flitting along alleyways, heading for the smoky warmth of their favourite tavern.

It would have been smoke, more than the warmth, that assailed the patrons of the Golden Grape as they entered the inn. The eye-watering fug that permeated every seventeenth-century tavern was so thick it was often difficult to see across a room. Part of it, certainly, came from the roaring open fires which were the only form of heating, but they were fuelled by local peat piled up in hollow pyramids in the grate. (Peat was excavated in such huge quantities that the Dutch of the Golden Age were creating new swamps and bogs almost as fast as they drained the old ones.) Visitors such as Peter Mundy found that Netherlands peat burned 'very Sweet and Cleare', even though the sulphur in it did turn those huddled around fires 'pale and livid, like ghosts'. So the smoke which filled the Golden Grape came almost entirely from the pipes smoked by its customers.

By 1636 pipe-smoking was so prevalent among the Dutch that it was practically a national characteristic. Tobacco, mostly imported from America but just now beginning to be grown in the United Provinces too, was taken in slim, long-stemmed clay pipes. Smokers puffed almost constantly, not least because the doctors of the period touted tobacco as a potent medicine, capable of protecting against the plague and curing everything from toothache to worms. The fact that tobacco was also said to soak up vital bodily fluids, making the men who smoked it infertile, does not seem to have put many people off. Entering the Golden Grape must have been like going into one of the overused and stale-smelling smoking rooms set aside by twentieth-century companies which have banned tobacco elsewhere in the workplace.

Once a newcomer's eyes had become somewhat accustomed to the murk, however, he would have seen that the tavern was packed and lively. Some of the details, which would not have struck a contemporary Haarlemmer as in any way unusual, might seem odd to modern eyes. One was the requirement to surrender weapons at the door, the result of one too many knife-fights in the past. (Dutchmen of the Golden Age had a dangerous passion for this sort of combat – 'a hundred Netherlanders, a hundred knives', as a contemporary proverb bluntly warned.) Another was the quality of the paintings displayed on the walls. Works of art were so ubiquitous in the Golden Age, and prices so low – a matter of a few stuivers or a guilder or two in some cases – that it was very common for taverns to display fine canvases or tapestries and allow them to yellow and blacken in the smoky air.

Most remarkable of all, though, was the sheer scale of the debauchery within. Even at a time when drinking was universal and drunkenness commonplace, the Dutch were Europe's most notorious sots. Beer was cheap – a whole evening's drinking could be enjoyed for less than a guilder – and Sir William Brereton found scarcely a sober man among the denizens of the Dutch taverns he visited. Even the English, no mean drinkers themselves, complained of the Hollanders' appetite for beer, and accused the Dutch of exporting the habit of drunkenness to Britain.

Virtually every Dutchman, in fact, frequented one tavern or another, as did many of the less genteel women and a good number of children. The atmosphere within these establishments was both convivial and inclusive, although there was a general suspicion, in many of the less salubrious establishments, that the staff were engaged in a systematic attempt to defraud their customers – which occasionally they were. As well as the usual tricks of short-changing sozzled patrons or watering down their beer, some inn-keepers coloured wine with sunflower or stuffed cloths into the bottom of their pitchers to reduce the amount of drink they would hold.

Visitors to such establishments were frequently appalled by the systematic way in which Netherlanders set about becoming intoxicated. Dutchmen seldom drank alone; they came in company, or would be welcomed into one of the groups already working their way through vast flagons of beer. Typically, the consumption of each new round would be prefaced by a toast, and this was one of the rituals which the tulip traders adopted with enthusiasm. 'These gentlemen,' the Frenchman Théophile de Viau observed of the habitués of one tavern he visited, 'have so many rules and ceremonies for getting drunk that I am repelled as much by the discipline as by the excess.'

It was, in any case, all but impossible to avoid beer in

the seventeenth century. The water was generally undrink-able – that would certainly have been true in Haarlem, thanks to the bleacheries – and tea and coffee were little-known luxuries; wine was relatively expensive. Beer was drunk with every meal: warmed and spiced with nutmeg and sugar at breakfast time, on its own at lunch and supper. Naturally not all the beer consumed in Haarlem was very alcoholic – it was brewed in two strengths, 'simple' and 'double', the former to quench thirst, the latter to intoxicate – but what there was was drunk in quantity. At the turn of the century, when the population of Haarlem was only 30,000 men, women, children and babes in arms, the con-sumption of beer ran at about 120,000 pints a day, which is five and a half million gallons a year, a third of which was drunk in taverns. To meet this demand, Haarlem alone contained about a hundred breweries, fifty of which were of a good size. The brewers were, in fact, not only wealthy but a potent political force in the city; a cabal of twenty-one of them had actually controlled the government of Haarlem for a number of years from 1618.

The florists of the city, cloistered in a back room of the Golden Grape away from the worst of the noise and smell of the city and the tavern itself, met by arrangement two or three times a week. In the early days of the tulip trade, these meetings took an hour or two at most, but as the mania took hold, the colleges began to sit for longer periods, sometimes starting in the morning and not concluding the last of their business until the early hours of the next day. Since each deal was celebrated with a call for wine – in itself a symbol of ostentation and wealth in what was a predominantly beer-drinking province – and since wine in

Dutch taverns was served in vast pewter pitchers which held anything from two pints to one and a half gallons, the trade was conducted for the most part in a haze of inebriation. Doubtless this, combined with the bravado generated by groups of friends, laughing and talking into the night, explains a good deal about the otherwise puzzling mechanisms of the mania.

In important respects, the tavern colleges seem to have operated independently from the rest of the tulip trade. They dealt in cheaper and more readily available bulbs, and their members, though they did include a few merchants and other affluent dealers, were drawn almost exclusively from the working classes. These men would have had little if any contact with the connoisseurs or established growers, and generally possessed at best second-hand knowledge not only of tulips, but also of finance, the stock exchange and the way that regents and great merchants dealt in shares and bought and sold commodities.

Many of the elaborate customs developed by the colleges seem to have been deliberately modelled on the methods of the stock exchange, a practice which must have heightened the florists' sense of self-importance and also helped to persuade the tulip traders that they were involved in a genuine and properly regulated business. Bulbs were put up for sale by auction, but where the more established growers and dealers sometimes visited a local solicitor and had their agreements notarized so as to ensure there was no possibility of any dispute, the florists substituted the quicker and cheaper system of recording all their transactions in their own bulky ledgers. Each college also elected a secretary who kept records of the deals struck around his table.

The tulips that were bought and sold by these tavern traders were rarely if ever the superbly fine varieties which obsessed the connoisseurs and wealthy dealers such as Abraham de Goyer. Probably they were bulbs of the second rank at first; and then, when demand rose further and even these became scarce, the colleges started to deal mostly in the least coveted and most common varieties of tulip. These flowers were known as *vodderij*, which means 'rags', or more politely as *gemeene goed*, 'common goods'. They were unicoloured or poorly variegated tulips which were often descendants of the earliest varieties to reach the United Provinces. Being both long-established and beneath the notice of the richer dealers, they were readily available by the end of 1636.

The *vodderij* were sold not by the ace but in baskets which were weighed out by the half-pound or the pound (a pound was 9728 aces in Haarlem, 10,240 in Amsterdam). In florists' slang they were often called pound-goods, to distinguish them from the piece-goods which were sold individually by the ace or by the thousand aces. A one-pound basket might contain as many as fifty or a hundred bulbs, and so a single tulip, even at the height of the mania, would have been priced within reach of all but the poorest traders.

The hundreds of novice florists who flocked to the bulb trade in the autumn and winter of 1636–7 generally began by dealing in small quantities of pound-goods, and the fantastic inflation that quickly occurred in the prices of these bulbs is a better indicator than any other of the vigour of the flower trade and the hold that tulip mania quickly exerted over the tavern colleges. A parcel of one of the cheapest pound-goods, Gheele Croonen, which could have been had for as little as 20 guilders in September or October 1636, cost 1200 guilders by the end of January. The more popular Switsers, still a comparatively dull variety, came on the market in the autumn of 1636 at 60 guilders per pound. But by 15 January 1637, the price was 120 guilders; on 23 January it was 385; and by 1 February it had all but quadrupled again, to 1400 guilders per pound. The peak price for this variety, recorded two days later, was 1500 guilders per pound.

Remarkable though the tulip's history had been up to this point, it was in the months of December 1636 and January 1637 that the bulb craze reached its peak and tulip trading turned into tulip mania. There are, unfortunately, no eye-witness accounts of what really went on in the tulip colleges during the extraordinary winter of 1636, or how exactly bulbs were bought and sold. However, the three *Samen-spraecken* appear to have been written by an author with a detailed knowledge of the tavern colleges, and they give a vivid picture of the mania at its height.

In the first pamphlet, Gaergoedt, the weaver who has become a florist, attempts to persuade his friend Waermondt to become a tulip dealer. He explains that he will teach him the secrets of the tavern trade and promises to tell his friend how to get himself admitted to one of the colleges and strike his first deal. Then he urges Waermondt to drink some wine with him. 'This trade,' he confides, 'must be done with an intoxicated head, and the bolder one is, the better.' As a one-line explanation of the worst excesses of the tulip mania, the weaver's aphorism could hardly be bettered.

First, Gaergoedt explains, Waermondt needs to find one of the taverns where the florists meet. There he should ask the landlord to show him into the presence of the tulip dealers. 'Because you are a newcomer,' he warns, 'some will quack like a duck. Some will say, "A new whore in the brothel", but take no notice.'

Once accepted into the company, the weaver continues, Waermondt can start to deal in bulbs. First he has to understand that it is rare for anyone to formally offer tulips for sale. Instead, florists are expected to make their intentions known by means of dropped hints and veiled allusions. It is, for example, permissible to say: 'I have more yellows than I can use, but I want some white.' When it does finally become clear there is a deal to be done, two methods of trading are employed in the tulip taverns, and which is used depends on whether one wishes to buy or sell. Either way, the trader chosen to be the secretary of the college will make a note of all transactions – and each and every deal means the donation of *wijnkoopsgeld* ('wine money') to the seller.

The first method, *met de Borden* ('with the boards'), was the one used by those who wished to buy. Wood-backed slates were given to both the buyer and the seller, and the florist who wished to buy would jot down the price he was prepared to pay on his slate; but he would choose a sum well below the actual value of the bulbs he wanted. The seller would name his own price on another slate, and naturally that would be exorbitantly high. The two bids would then be passed to intermediaries nominated by the principals, and they would mutually agree on what they considered a fair price. This sum would fall somewhere between the two prices written on the slates, but certainly not necessarily in the middle. The compromise price would then be scrawled on the slates and the boards would be passed back to the florists.

At this point, bulb buyer and bulb seller had the option of either accepting or rejecting the arbitration. They accepted by letting the revised price stand; at that point the transaction was concluded and the purchase price would be noted in the college register. The buyer was then expected to pay a commission of half a stuiver per guilder of the purchase price; if the agreed price was 120 guilders or more, the commission remained fixed at the maximum of 3 guilders. This was the *wijnkoopsgeld*. If, however, either the buyer or the seller did not want the deal to go through, he could signal his refusal to accept the compromise price by rubbing it off his slate. If both parties did this, the deal was off; but if only one rubbed out the new price, he had to pay a fine of somewhere between 2 and 6 stuivers for his intransigence. Thus the *met de Borden* system did offer an incentive to trade.

Those who wished to initiate a sale employed a slightly different system known as *in het ootje*, 'in the little "o"'. Today this phrase is a piece of Dutch slang which means to pull someone's leg, but during the tulip mania it referred to a portion of the rough diagram that the secretary of the college would draw to keep track of the bidding in what was effectively a form of auction. The diagram looked like this:

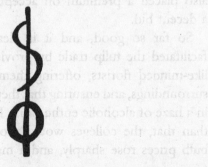

When selling *in het ootje*, this same figure was sketched on the slates of each member of the college. A florist who wished to dispose of some bulbs would write in the small 'o' at the bottom of the diagram the number of stuivers he was prepared to donate as a bounty or commission to a buyer. The amount would vary depending on the seller's assessment of the value of his bulbs, but again it would be somewhere between 2 and 6 stuivers, that is, about the cost of a round or two of drinks. Prospective florists among the college would then offer what they thought the tulips were worth, the secretary keeping track of the bids by noting down the highest offer in thousands in the top semicircle, in hundreds in the bottom one, and in units underneath the vertical line. When the bidding was at an end, the secretary would strike three lines through the diagram on his board and surround the whole thing with a big 'O' – the tulip trade's equivalent, it would appear, of the modern auctioneer's cry of 'Going, going, gone'. This concluded the auction, and the seller had the option of accepting or rejecting the highest bid; but if he refused it he still had to give the thwarted buyer the commission specified *in het ootje*. This method of trading bulbs, then, also placed a premium on accepting rather than rejecting a decent bid.

So far so good, and it is clear that the tavern clubs facilitated the tulip trade by providing a meeting place for like-minded florists, offering them warm and comfortable surroundings, and ensuring that their business was conducted in a haze of alcoholic enthusiasm. If they had done no more than that, the colleges would probably have ensured that bulb prices rose sharply, and a mania of some sort would have ensued. In fact, the customs of the tavern trade had an even greater impact.

First, as we have seen, the colleges proved willing to trade not just real, physical tulips, but also the rights to ownership of bulbs which were still in the ground. Thus they changed the tulip trade from a seasonal thing, possible only for a few summer months after the bulbs had been lifted, to a business which could continue all the year round. This gave the traders – who, it must be remembered, rarely had gardens of their own to tend – something to do during the winter, maximized their potential for profit, and also ensured that the *wijnkoopsgeld* continued to flow to everyone's satisfaction. Second, the colleges failed utterly to check whether their members had enough money to cover their debts, or even owned the tulips they traded. In the absence of physically real bulbs, this would seem to be an elementary precaution, but they did not take it. The tavern clubs thus encouraged unbridled speculation, while offering their members absolutely no safeguards against insolvency and fraud. It was now quite possible for a florist who owned no bulbs to begin trading, in the expectation that he would be able to shift his obligation to actually buy a given bulb on to another dealer long before he was called to account, and then use the profit on that deal to fund his next purchase. And it was equally possible for the same man to become technically insolvent the moment the price of tulips fell.

In the *Samenspraecken*, Gaergoedt boasted of earning 60,000 guilders from the flower trade in only four months. In the winter of 1636–7, the real tulip maniacs were to get the chance to see if they could match him.

CHAPTER 11

The Orphans of
Wouter Winkel

Tulip mania had made Wouter Bartelmiesz. Winkel one of
the richest men in the town of Alkmaar. Although a mere
tavern-keeper by trade (he was the landlord of an inn called
the Oude Schutters-Doelen – 'The Old Civic-Guardhouse' –
in the centre of the town), he could count on the fingers
of one hand the number of fellow citizens who were
wealthier than he. The only problem, which he shared with
every other tulip dealer, was that he could not lay his hands
on his money. It lay buried in the ground in the form of
bulbs.

Wouter Bartelmiesz. seems to have come originally from
the village of Winkel, which lies about 10 miles to the north
of Alkmaar in the furthest tip of the province of Holland.
His parents, while not wealthy, appear to have been reasonably well off. His brother Lauris was able to complete
an apprenticeship and become a goldsmith, which was
always one of the best-paid occupations that a member of
the artisan class could aspire to. And when Wouter married
Elisabet Harmans in 1621, he promised his wife that they

could afford a large family of their own. No fewer than seven of the children he had with Elisabet survived infancy, and because even in 1636 only one, fourteen-year-old Willem, was old enough to start earning his own living, the whole family must have been supported by the profits of the tavern and Winkel's bulb trade.

Alkmaar was one of the smaller towns of the United Provinces, but to a villager from Winkel it must have held all the allure of a metropolis. It was the market town for much of what was called the North Quarter of Holland, where it competed with its ancient rivals Hoorn and Enkhuizen for trade, and it was a notoriously independent place, uninterested in conforming to the fashions of the rest of the Republic. The women of Alkmaar, for example, almost alone among the Dutch, did not wear white linen caps, but fashioned their hair into an extraordinary style – all interwoven braids – which resembled a sort of helmet.

The expanse of countryside that the town dominated had shrunken considerably since the Middle Ages, when it had effectively controlled most of north Holland and even several of the islands strung across the mouth of the Zuyder Zee, but it was still surrounded by rich farmland and had benefited considerably from the recent draining of some of the small lakes to the south. The town specialized in beef and dairy produce, and particularly in the huge wheel-shaped cheeses that had already made the United Provinces famous throughout Europe.

The Winkel family seems to have prospered in Alkmaar for a while, but like every other family of the period, they lived lives that were permanently on the brink of tragedy.

Even during its Golden Age, the Dutch Republic remained prey to many of the dangers that made life in seventeenth-century Europe so frequently miserable. It was an era of war and want, low life expectancy, recurrent plague and high infant mortality; the few doctors were still all but helpless in the face of even common illnesses, and the potions and cures they did prescribe were frequently more deadly than the ailments they were supposed to counter. Few families could hope to go through life without losing a child or two, a husband or a wife.

In the Winkel family it was Elizabet Harmans who went first. She died some time between 1631 and 1635, perhaps of disease, perhaps in childbirth, leaving her husband with three boys and four young girls to care for. There is no record of a second marriage, so the presumption is that Winkel struggled on very much alone, his older children helping to take care of their younger brothers and sisters, perhaps with the assistance of a servant or the serving girls at the Oude Schutters-Doelen.

In those days Dutch children began their schooling at the age of seven, so the whole family except the youngest, a boy of six named Claes, were already of school age. That suggests that Wouter Winkel would not necessarily have had to hire anyone to help him with the children. Even so, he would undoubtedly have felt the loss of his wife financially as well as emotionally. Someone would have to be paid to do the sewing, the cleaning and the cooking that Elizabet had done, and so the profits of the tulip trade would have been even more important to the surviving members of the family now.

Wouter Bartelmiesz. seems to have got involved in bulb

dealing relatively early on. He was certainly buying and selling tulips in 1635, well over a year before the market really boomed, and the chances are that he started dealing in bulbs a year or two before that. This early start, combined with a little luck and a good understanding of the flower trade, enabled him to amass a tulip collection of quite spectacular quality.

By the spring of 1636, the tavern-keeper owned more than seventy fine or superbly fine tulips, representing about forty different varieties, together with a substantial quantity of piece-goods totalling about 30,000 aces of lower-value bulbs. His tulips included some of the most valuable flowers to be found anywhere in the United Provinces: a very rare *Violetten* called Admirael van Enkhuizen, together with two Viceroys and five Brabansons of various types; three bulbs of the celebrated *Rosen* Admirael van der Eijck, an Admirael Liefkens, a Bruyn Purper ('Brown and Purple'), a Paragon Schilder and no fewer than seven examples of the increasingly sought-after Gouda. At the height of the mania, bulbs of every one of these varieties could easily change hands for 1000 guilders, and often substantially more. Assembling such a quantity of the most sought-after tulips in the United Provinces was an astonishing feat of dealing on Winkel's part. If his tulips were not the most fabulous collection of flowers in the Republic, they must have come close, for no other record has yet been found of a bulb trader whose tulips even approached the quality and variety of those owned by Wouter Bartelmiesz.

The most impressive thing about Winkel's collection, though, was neither the variety nor the magnificence of the tulips in it, but the fact that he actually owned every flower in his inventory. Wouter might have been a tulip trader, but he was neither a connoisseur nor a florist; he was a grower. That meant his assets were more substantial than those of the majority of dealers, who owned nothing but promissory notes inscribed with a price and a notional delivery date, and had no guarantee that their tulips were of good quality, or even that they actually existed. Winkel's assets were bulbs, planted in a garden close to his inn.

Unfortunately for Wouter Winkel and his seven children, he did not live long enough to reap the enormous profits that his canny trading would have earned him. He saw his tulips flower in the spring of 1636, but died some time in the early summer, probably aged only in his late thirties or early forties. We do not know what accident or illness killed him, only that shortly afterwards, a party of grim-faced representatives of the local Orphans' Court arrived at the Oude Schutters-Doelen and took the tavern-keeper's children off to Alkmaar's orphanage.

In some respects, the children's plight was not quite as catastrophic as it appeared. The death of both parents was a relatively common occurrence in the seventeenth century, and the United Provinces probably made better provision for caring for its orphans than any other country at that time. Most places of any size had their own orphanage, funded by the town and governed by a board of regents who assumed responsibility for the childrens' interests, supervised the full-time staff and made sure sufficient funds were raised to keep the institution running smoothly. The same cities typically also ran homes for the elderly – one for men and another for women – which were open to any aged citizens who met certain residency requirements. These early social services were unique to the Dutch and were the envy of the foreigners who saw them.

Nevertheless, the orphans of Wouter Winkel faced an uncertain future if they stayed in the Alkmaar orphanage. Their guardians, their uncles Lauris Bartelmiesz. and Philip de Klerck, would no doubt do what they could to help them, and the town would feed and clothe and school them for a year or two. But they were assured of board and lodging at the orphanage only until they were old enough to work for their living. Then they would be packed off to some factory, mill or workshop to learn a useful trade which would ensure they did not remain a burden on their home town. The children would have little choice as to where they were sent, though they might then be no worse off than the children of other artisans. Still, these orphans had only one chance of assuring themselves a more comfortable life: they had to sell their father's flowers.

The first step was to ensure that the tulips were safe. This was a very necessary precaution. As prices spiralled ever upwards, every grower feared the loss of his bulbs, and some were already taking elaborate precautions to guard them. Some slept with their tulips and one man, from the village of Blokker, installed trip wires around his bulbs and connected them to a bell which hung close to his bed. Confined to their orphanage, and with their father dead, the Winkel children's bulbs would have been especially at risk, and they must have greeted lifting time with some relief. Within a day or two all the bulbs had been safely gathered and locked in a secure room in the orphanage while the Trustees of the Orphans' Court considered how best to proceed.

This was in July 1636. It was not until December, however, with the bulbs carefully graded and weighed and back in the ground once more under the watchful eye of a gardener named Pieter Willemsz., that the Trustees finally authorized a sale.

It is not clear whether this long delay was caused by the labyrinthine bureaucracy of the Orphans' Court or whether one of the regents of the orphanage had watched the rise in tulip prices and waited for the right moment to sell the Winkel bulbs. But whether it was by accident or design, the auction which finally took place at the Nieuwe Schutters-Doelen, Alkmaar, on 5 February 1637 was held at the perfect moment. In the months since Wouter Winkel's death, tulip prices had doubled, then doubled and doubled again. With so many new buyers in the market, his rarest and finest bulbs were now far more sought after than ever before.

The Trustees of the Orphans' Court had taken care to publicize the sale, and the innkeepers of Alkmaar must have done good business as dozens of wealthy florists and growers crowded into the town in the first few days of February. Potential bidders were invited to inspect a special tulip book commissioned by the Court which contained 124 watercolours of Winkel's tulips and 44 of the lilies, anemones and carnations which made up the balance of his collection. The book acted as a sort of sale catalogue and a promise to potential buyers of the glories that could be theirs in only a month or two if they bid successfully.

The auction at Alkmaar was the supreme moment of the tulip mania. The crowd attracted to the sale seems to have been a cut or two above hoi polloi of the taverns, and almost certainly the bidders would not have been permitted to get away with college practices such as offering just part payment in kind. This was an auction for connoisseurs and affluent dealers. Real bulbs were being sold on a large scale for cash.

Even before the proceedings began, one determined buyer had contrived to negotiate privately with the regents of the orphanage for the jewel of Winkel's collection, the *Violetten* Admirael van Enkhuizen. When this tulip had been lifted the previous summer, the mother bulb was found to have grown a small offset which promised to become a viable bulb itself in the new year. The presence of this offset substantially increased the value of the already rare bulb, and the regents sold it for an astonishing 5200 guilders, close to the price that had been quoted for a Semper Augustus in 1636. The same wealthy buyer also purchased two of the increasingly popular lilac-flamed Brabansons for 3200 guilders the pair, and a miscellaneous lot which appears to have consisted of some more rare tulips and Winkel's collection of lilies, carnations and anemones. For these flowers the buyer paid an additional 12,467 guilders – a staggering total, for this one sale alone, of more than 21,000 guilders, was enough to buy not one but two large houses on the Kaizersgracht in Amsterdam.

The lucrative private sale of these few bulbs set the tone for the auction which now began. The buyers appear to have been convinced, either by the tulip book or by Winkel's reputation, that the flowers were of the highest quality and that this was a rare opportunity to acquire some of the most sought-after tulips in the United Provinces. They bid fiercely, and the prices achieved at Alkmaar were, with few excep-

tions, the highest ever recorded for the various tulips on
sale.

Most of the best lots were concentrated at the beginning
of the auction. The first, a 563-ace bulb of a middle-ranking
red and white variety called Boterman, sold for 263 guilders,
about half a guilder per ace, but the next, a tiny Scipio of
only 82 aces, was knocked down for 400 guilders – 5 guilders
an ace. A Paragon van Delft sold for 605 guilders, then
Winkel's prized Bruyn Purper, a subtle flower which mixed
a hint of brown into its lilac flares, went for 2025, which
was 6 guilders 7 stuivers per ace.

So it went on, bulb after bulb attaining record prices. Only
two of the seventy main lots which went under the hammer
in the first part of the auction sold for less than 100 guilders,
and nineteen tulips were valued at more than 1000 guilders
each. The most expensive bulbs were two good-sized Viceroys
of 658 and 410 aces, which sold for 4203 and 3000 guilders
respectively, but in terms of value per ace the most coveted
flower was a *Rosen* of the variety Admirael Liefkens. This bulb,
when planted, weighed a mere 59 aces, which made it the
lightest tulip (bar one) to be sold that day. It can have been
little more than an offset, but it cost its buyer 1015 guilders,
which is 17 guilders 4 stuivers per ace.

Even the cheaper piece-goods which were sold at the end
of the day, after all the superbly fine tulips had found
buyers, attained good prices. Five hundred aces of *Violetten*
Rotgans were sold for 805 guilders to one dealer and for
725 to another, and 1000 aces-worth of bulbs produced by
Jan Casteleijn, a Haarlem grower who had a garden on the
south side of the Campeslaen, were knocked down for 1000
guilders.

Even before the auction had finally drawn to an end, it must have been obvious to those watching the bidding that Wouter Winkel's bulbs were commanding sums that were staggering even by the standards of the tulip mania. In addition to the 21,467 guilders raised in the earlier private sale, the seventy individual tulips auctioned at the Nieuwe Schutters-Doelen were sold for a combined total of 52,923 guilders and bulbs of the twenty-two varieties sold by the 1000 aces for a further 15,610. The total for the whole auction and the private sale combined came to a round 90,000 guilders.

In the space of an hour or two, the Winkel children had gone from being poor orphans to exceptionally rich young men and women. We know nothing about how the money raised at the auction was collected and what commissions, deductions and tax may have been payable on this fantastic windfall. But if each of Winkel's seven offspring simply took one-seventh of the gross sum, they would have received almost 13,000 guilders each, more than forty times the annual income of a typical artisan family. An ambitious boy could take that money and buy his way into almost any profession he cared to enter. A cautious one, living modestly, need never do a day's work in his life. And any girl with such a dowry could rely on making a very advantageous match.

There does not seem to have been any doubt in the mind of Dutch tulip traders that the auction at Alkmaar was an extraordinary event which deserved to be commemorated. Within a few days of the sale, a one-page pamphlet with the modest title *List of Some Tulips Sold to the Highest Bidder on February 5th 1637* had gone on sale. It gave a few brief details of the circumstances of the auction and listed the prices paid for each of the ninety-nine lots. Some writers have suggested that this pamphlet was meant as a warning against extravagance, but its principal purpose would appear to have been to boost confidence in the tulip trade by making as many people as possible aware of the phenomenal prices that bulbs were now commanding.

In this it was at least partially successful. The pamphlet attained a sufficiently wide circulation for the prices it listed to be regarded as in some way official, even typical. A number of contemporary tulip books which actually indicate the cost of various varieties give the prices attained at Alkmaar, even though they were far in excess of anything that had been realized earlier in the mania. (The idea, presumably, was to persuade potential purchasers that they should pay high prices.) Thus Admirael Liefkens, the most expensive tulip purchased at the auction in terms of gilders paid per ace, had been worth only 6 guilders, 12 stuivers per ace in June 1636, and Winkel's three Admirael van der Eijcks – bulbs of a variety that had sold at 2 guilders, 10 stuivers per ace the previous July – went for as much as 7 guilders, 14 stuivers per ace at Alkmaar.

Tulip mania reached its peak throughout the United Provinces in the last week of January and the first week of February 1637. During this extraordinary fortnight, huge amounts of money were pledged in moments. Hendrick Pietersz., a baker from Haarlem, paid 100 guilders for a Gouda weighing just seven aces – a price of more than 14 guilders per ace, one of the highest ever recorded for a tulip. And extracts from the trading ledger of a Haarlem merchant named Bartholomeus van Gennep, preserved in the legal archives of his city, show that late in January he agreed to pay a single dealer, Abraham Versluys, more than 3200 guilders for a collection of second-rate bulbs which included none of the highly coveted varieties most commonly associated with the mania:

Two pounds of Yellow and Red Crowns	385 guilders
A pound of Switsers	280 guilders
3000 aces of Centen	380 guilders
Half a pound of Oudenaers	1430 guilders
1000 aces of Le Grands	480 guilders
1000 aces of Gevleugelde Coornharts	220 guilders
70 aces of Kistemaecker	12 guilders
410 aces of Gevlamde Nieulant	54 guilders
	3241 guilders

Although the passion for bulb dealing was still concentrated in its oldest strongholds, Haarlem and Amsterdam, it had now reached beyond the borders of Holland and West Friesland – certainly to Utrecht and Groningen, and most probably to other provinces as well. Indeed, the horticulturalist Abraham Munting (who was a boy during the mania) noted, without giving details, that speculation in tulips was raging, for a second time, in northern France as well.

The number of people involved in buying and selling tulips across the United Provinces must by now have been in the thousands. One of the few detailed documents that have survived suggests that in the city of Utrecht, which was far from being one of the largest centres of the bulb trade, there were about forty serious growers in February 1637. This almost certainly means that a couple of hundred florists and hangers-on also traded in the town. Since bulb growing and flower dealing flourished in at least a dozen cities and districts in Holland and neighbouring West Friesland alone, from Medemblik in the north to Gouda in the south, it is probably safe to estimate that a minimum of three thousand people found themselves caught up in the tulip mania in the central Netherlands. If so, there could scarcely have been fewer than five thousand growers and florists among the two million citizens of the Dutch Republic as a whole by the time the mania reached its peak, and this figure may well be a very conservative estimate.

The total value of the flowers bought and sold by such a large number of people must have been staggering. Some authorities suggest that at the height of the mania, bulbs were changing hands as often as ten times in a single day – the price rising, presumably, with every deal. Thus, while the tulips themselves remained securely in the ground, they might be owned by a weaver and a glass-blower, a fuller and a clerk, all in the space of twenty-four hours, and emerge from the soil at lifting-time worth five or ten times more than they had been when they were planted. The rarest bulbs, on the evidence of the Alkmaar auction, could total 4000 or 5000 guilders apiece, and even if we accept that the prices realized at the Nieuwe Schutters-Doelen auction were exceptional, it certainly does not appear to have been unusual for superbly fine tulips to have changed hands for a couple of thousand guilders apiece, lesser varieties for somewhere between the 350 guilders per 1000 aces recorded for the red-and-white Centen and the 750 paid for the popular *Bizarden* Root en Gheel van Leyde, and mere pound-goods for somewhere between 250 guilders and 1500 guilders a pound. Thus, even if we make the conservative estimate that in one of the largest centres of the tulip trade, such as Haarlem or Amsterdam, a total of say four hundred florists met in colleges to deal in bulbs four times a week, the amount of money which changed hands in that one city during the three or four months that the mania was at its peak could have run well into seven figures. If, for example, the typical florist bought just one pound of tulips a day at an average price of 250 guilders per pound, the volume of trade in a single large town would have approached 7 million guilders between the beginning of October 1636 and the end of January 1637 alone.

Some dealers appear to have been at least this active. While the mania was at its peak in December and January, a single tulip trader, Pieter van Rosven of Haarlem, bought bulbs worth 2913 guilders in the space of only six weeks, mostly from Wouter Tulckens of Alkmaar. Tulckens appears to have acted as a sort of broker for several growers. One of the bulbs he sold to van Rosven was planted in the garden of a Cornelis Verwer, another in a plot kept by the Calvinist minister Henricus Swalmius on the Bollslaen ('Bulb-lane') which lay to the south of Haarlem, and a third in the garden of the painter Frans Grebber. And these are merely the purchases that found their way into the legal archives of Alkmaar as a result of an action van Rosven brought against Tulckens for the nondelivery of his bulbs; he may well have bought and sold many other tulips in the same brief period.

Van Rosven's example was certainly not unique. In the *Samenspraecken*, Gaergoedt relates that in the colleges he attended, so many bulbs were bought and sold at high prices that *drietjens* – the maximum 3-guilder sums paid as wine money when a deal exceeded 120 guilders – 'fell like drops of water from thatched roofs when it has rained'. 'I have often been to inns and eaten baked and fried fish and meat,' the florist adds. 'Yes, chickens and rabbits, and even fine pastry, and drunk wine and beer from morning to three or four o'clock at night, and then arrived home with more money than when I left.' When the chronicler Lieuwe van Aitzema guessed that 10 million guilders' worth of tulips changed hands in a single Dutch town in the course of the mania as a whole, then, he may actually have been underestimating the true extent of the trading frenzy.

It is hard to resist the conclusion that in crude financial terms the mania really was an event of unparalleled magnitude, at least by the standards of the United Provinces. If we accept that van Aitzema was right, and that perhaps 20 million guilders' worth of bulbs were bought and sold in Haarlem and Amsterdam together between 1633 and 1637, then, even supposing that the trade in each of the other ten known centres of the mania amounted to no more than a tenth of that in those two great centres, the nominal turnover of the Dutch bulb trade as a whole in those four years can hardly have been less than 40 million guilders. And if the florists of Holland really did trade as compulsively and as irresponsibly as the critics of the bulb trade maintained, if the number of people involved was not thousands but tens of thousands, then the total could conceivably have been twice that figure or even more. By way of comparison, the total sum deposited by rich merchants with accounts at the Bank of Amsterdam in the years 1636–7 was probably only about 3.5 million guilders, and the all-powerful Dutch East India Company – which was the greatest trading organization in the whole of Europe at the time – was capitalized at 6.5 million guilders.

It was left to a contemporary pamphleteer, writing around December 1636, to give perhaps the most vivid impression of what the prices paid for tulip bulbs actually meant to the Dutchmen of the time. A flower worth 3000 guilders, the writer pointed out, could have been exchanged for a gigantic quantity of goods:

Eight fat pigs	240 guilders
Four fat oxen	480 guilders
Twelve fat sheep	120 guilders
Twenty-four tons of wheat	448 guilders
Forty-eight tons of rye	558 guilders
Two hogsheads of wine	70 guilders
Four barrels of 8-guilder beer	32 guilders
Two tons of butter	192 guilders
A thousand pounds of cheese	120 guilders
A silver drinking cup	60 guilders
A pack of clothes	80 guilders
A bed with mattress and bedding	100 guilders
A ship	500 guilders
	3000 guilders

From this perspective, it is evident that the tulip trade really was not merely healthy but positively booming in

Holland in the autumn and winter of 1636–7. Yet even as the mania reached its peak, there were disturbing indications that all was not well in the tavern colleges.

One warning sign was the florists' restless search for novelty. Although most seem to have agreed that one or two varieties, such as Viceroy, stood apart from their many rivals, there was very little consensus on which tulips deserved to stand in the second rank, and this problem was exacerbated by the similarities between many of the more popular flowers; even an expert would find it difficult to tell an Admirael van Engeland from an Admirael van Hoorn. One college or town favoured one variety, others another; furthermore, fashions and opinions changed, and new tulips continued to come on the market to challenge the established favourites. Because of this, the trade in bulbs was not just unstable, but inherently illogical. No market can flourish for long if it does not possess elements of stability and predictability. The Dutch tulip trade had neither.

A notorious example of the florists' restless desire to find something new and different was the quest for the black tulip, a flower of such fabled rarity that it would certainly have been worth considerably more than even Semper Augustus if only an exemplar could have been found. Although the French novelist Alexandre Dumas turned the black tulip into a famous novel, which deals with the efforts of a young physician named Cornelis van Baerle to win the huge prize offered to the first man to grow such a flower, he seems to have chanced upon the idea in an old Dutch account of an incident which is supposed to have taken place at the height of the mania. According to one version of this story, a syndicate of Haarlem florists heard that a cobbler living at The Hague had succeeded in breeding a black tulip and determined to purchase the solitary flower that he possessed. They visited him in his shop and, after a certain amount of bargaining, the cobbler agreed to accept 1500 guilders for his tulip and handed over the bulb. To his complete astonishment, the Haarlem florists immediately hurled the black tulip to the ground and trampled on it, exclaiming: 'Idiot! We have a black tulip too, and chance will never favour you again. We would have given you 10,000 guilders if you had asked it.' Assured that their own bulb was once again unique, and thus beyond price, the Haarlemmers returned to their city, leaving the unfortunate cobbler so distraught at the thought of the wealth that might have been his that he died of nervous shock the same night.

The story of the black tulip is, of course, a myth. Indeed the botany of the genus is such that it is actually impossible to breed a flower with pure black petals; even today, the handful of 'black' tulips which exist are merely an exceptionally dark shade of purple. Nevertheless, the fact that the black tulip legend achieved a certain currency during the years of the tulip mania might perhaps have alerted the more astute florists to the fact that a dangerous gap was beginning to develop between what the bulb market was demanding and what was actually possible given the time it took to introduce a new variety and the limited stock of botanical tulips which Dutch growers had to work with.

Still more worrying, in the autumn of 1636, was the boom in the market for pound-goods. The astonishing prices being asked for these previously worthless bulbs by the beginning of 1637 should have given any florist pause for thought. However high the prices of superbly fine tulips rose, there was always some justification for them. Throughout the mania years, there remained a small but genuine demand for such bulbs from the old tulip connoisseurs – the only people who actually wanted to plant and grow the bulbs rather than simply trade them. No such demand existed in the case of pound-goods. Connoisseurs would not touch them and most college florists were not remotely interested in cultivating them. They were traded solely because they were available, and as January gave way to February, even devoted tulip maniacs began to be uneasily aware that their market was running out of control.

The success of the auction at Alkmaar did provide some reassurance that high prices were still being paid for bulbs, but, even so, a few of the more cautious dealers must have begun to wonder just how much longer tulip prices could continue to rise. Here and there an isolated florist sold his holdings and declined to reinvest his profits in yet more bulbs. In the tavern colleges dotted across Holland, rival traders looked on and wondered if the seller knew something they did not. Perhaps, they thought, they might dispose of a bulb or two themselves.

It was the first week of February 1637. The boom time was over.

CHAPTER 12

Bust

The great crash in tulip prices began in Haarlem on the first Tuesday in the month, when a group of florists gathered to buy and sell as usual in one of the city's tavern colleges.

As was customary, an established member of the college began the day's trading by testing the state of the market; he offered a pound of Witte Croonen or Switsers for sale. The florist asked a fair price for the bulbs, 1250 guilders, and in the normal course of events he would have found several eager buyers. Slates and chalk would have been distributed, the tulips would have been knocked down to the highest bidder, and the rest of the day's trading would have continued in its usual frenzied way. On this day, however, there were no bidders for the bulbs at 1250 guilders. The auctioneer offered them again, this time cutting the price to 1100 guilders. Still there was no interest. Desperately now, he offered his bulbs for a third time, dropping his price to a risible 1000 guilders the pound. Once again there were no bids.

It is easy to imagine the awkward silence that must have

descended upon the group of florists hunched around their table in the tavern as this awful, farcical auction ran its course: the half-drunk tankards of beer hanging in mid-air, partway to the lips of drinkers who suddenly understood the importance of what was happening in front of them; the nervous glances exchanged by traders who had not the slightest idea what to do next or how they were supposed to react. The silence must have lingered for a second or two and then been broken by a swelling hubbub as every anxious florist present started talking at once.

In all likelihood, every one of the traders present had paid similar prices for similar bulbs within the past few days in anticipation of selling again for another handsome profit. Now, in a matter of no more than a minute or so, their assumptions had been shattered and the ugly question of what would become of the bulb trade had been raised. Plainly it would have been impossible to continue with the normal business of the day. Perhaps some brief attempt was made to sell other bulbs, without success, but the college must have suspended trading almost immediately, and in the general confusion one or two of the assembled company no doubt ran to inform their friends and family of what had occurred. In no time all the colleges in Haarlem would have heard the news, and every florist in the town and beyond the walls would have been gripped by a simple impulse: sell.

It took only a few days for the panic to spread through the rest of the United Provinces. In college after college, and in town after town, florists discovered that flowers which had been worth thousands of guilders only a day or two before now could not be sold for any price. A few dealers kept their heads and tried to stimulate renewed interest by organizing mock auctions or offering bulbs at huge discounts, but they were ignored. In most places, the tavern trade crashed so completely that it was not even a question of prices falling to a quarter or a tenth of what they had been when the mania was at its peak. The market for tulips simply ceased to exist.

Many florists must have found themselves in the same plight as Gaergoedt, the weaver described by the author of the *Samenspraecken*. Caught out by the unexpected fall in prices, Gaergoedt's first reaction is to go out to buy and sell again. He retains some of his old confidence, telling his friend Waermondt: 'Flora may be ill, but she will not die.' While his wife Christijntje bewails her husband's decision to sell off his loom and all his weaving tools, Gaergoedt returns to the colleges, only to find that the market really has collapsed and all trading has come to a halt. Unable to find a single buyer for his bulbs, and conscious of the many debts he has incurred in buying tulips and laying out a garden, the hapless weaver asks his friend what he should do. Waermondt's advice is brutally straightforward: the tulip trade is dead, he says, there is no chance of reviving it, and the florists have no option but to return to their old jobs and their proper stations. The best they can hope for is to be given the opportunity to discharge their debts honourably.

Confidence is everything in a booming market, but the fact that the tulip trade collapsed so rapidly suggests that some of the less sanguine florists must have been feeling uneasy about the continual escalation of bulb prices for a few days before the crash. The mania took place before the introduction of daily newspapers, so there is no way of being certain of the sequence of events in the final week of January and the first few days of February, but it is unlikely that bulb dealing ceased without prior warning and with complete finality simply because of one failed auction in a single Haarlem college. Trading must surely have become more and more difficult everywhere in Holland over the previous week or so; auctioneers would have found it harder and harder to push prices up at the old rapid rate, some varieties would have peaked in value, and the number of dealers anxious to sell would have begun to outnumber those still willing to buy. In the day or two before the fateful meeting at Haarlem it is not too fanciful to suppose that a general feeling of unease and trepidation must have descended upon the colleges of Haarlem and Amsterdam like a clammy autumn fog rolling in off the Zuyder Zee. The tulip traders had been waiting for something to happen, and now it had.

Certainly rumours that the markets could rise no higher were already in the air before the fateful meeting at Haarlem on 3 February, and some buyers were no longer confident that their investments would yield a certain profit. As early as the end of December, an apothecary and bulb grower called Henricus Munting, who lived in the town of Groningen, was able to complete a lucrative deal to sell a handful of his tulips for 7000 guilders to a man from Alkmaar only by promising his nervous customer that if prices fell before the summer of 1637 he could cancel the purchase and pay no more than 10 per cent of the agreed price. Then, two days before the crash, at a dinner party held at Pieter Wynants's house in Haarlem, Pieter's younger brother Henrik tried to badger one of the guests into buying a pound of Switsers for 1350 guilders. Henrik's target was a rich widow named Geertruyt Schoudt, but she demurred and could not be persuaded to buy. It was only when another dinner guest, a local dyer named Jacob de Block, offered to guarantee the price for eight days that Schoudt gave way and bought the bulbs.

The collapse of the tulip trade after 3 February was so complete that virtually no information has survived concerning the sort of prices paid for bulbs in the spring of 1637. It would appear that the only buyers left in the market were the connoisseurs and perhaps a few rich florists who were not entirely dependent on flowers for their wealth, and that only the rarest and most superbly fine varieties had any chance at all of being sold. According to one contemporary, a tulip that had been worth 5000 guilders before the crash was sold later for only 50 guilders. In May, a bed of tulips which would have fetched 600 to 1000 guilders in January is said to have changed hands for 6 guilders, and a selection of bulbs worth about 400 guilders during the boom was sold for a mere 22 guilders, 1 stuiver. These prices suggest that where tulips could be sold at all, they fetched, at best, just over 5 per cent of their old values, and often 1 per cent or less.

This was a truly spectacular crash. Even if it was not more or less instantaneous in each of the towns affected by the mania – and it probably was – the collapse of the tulip trade certainly cannot have taken more than three or four months. It was thus far more rapid and far more complete than history's most notorious financial disaster, the Wall Street Crash of 1929 and the Great Depression which followed it. In that case it took more than two years for share prices to fall to their lowest point, and even then they retained 20 per cent of their old value.

Amid all the confusion, few florists seem to have understood exactly why the bulb trade had collapsed in such spectacular fashion. Yet in retrospect it is not difficult to see that the crash was all but inevitable. Like a sun, tulip mania burned brightly and steadily while there was still fuel to feed it in the shape of a steady supply of bulbs. But during the winter of 1636–7, demand for tulips comprehensively outstripped supply and the mania then began, in effect, to consume everything around it. Pound-goods and unicoloured tulips were pressed into circulation, and in a market where even the hitherto despised Switsers and Witte Croonen were selling for more than 1000 guilders a pound, the florists of Holland were dealing every last bulb they could lay their hands on.

Once even these *vodderij* were being traded, there were no longer any new varieties coming on the market at affordable prices. The absence of cheap bulbs meant, in turn, that it was all but impossible for more novice florists to enter the market, for who could afford to do so if the very cheapest lots were selling for dozens or even hundreds of guilders? A handful of the existing traders were even selling up and trying to take their profits, so a shrinking group of florists, possessed of only a limited amount of capital, was somehow sustaining a constant rapid rise in prices. Sooner or later, even those who still believed the trade was fundamentally healthy would become unable to afford the next price rise and hesitate to commit themselves. Thus, by the beginning of February, money and bulbs – the twin fuels of the flower mania – were both exhausted. And like a sun which has burned the last of its fuel, the tulip mania 'went supernova' in a final, frenzied burst of trading before collapsing in on itself.

That was the reason for the crash, but not for the sheer extent of the collapse in prices. The explanation for that catastrophe lies in the extraordinary rapidity with which bulbs had passed from one hand to another at the height of the boom. In most bull markets there are also bears, who hold back capital and wait for prices to fall so they can buy valuable stocks cheaply. But the majority of the tulips which were traded in the last month or two of the mania – the pound-goods and some of the varieties sold by the 1000 aces – were literally worthless. There was no demand for them, no connoisseur would plant them, and they had value only in the eyes of the people who had traded them. There was nothing there for a bear trader to exploit.

Worse, it would appear that tulip mania had sucked in everyone it touched in the tavern colleges. Few florists had had much capital to spare when they entered the market, and nearly all were now caught in one or more of the complex chains of obligation which the bulb trade had created. A large number had sold or mortgaged their few possessions to finance their dealings in the bulb market. Those who were in this desperate position faced not merely loss, but ruin; and in the seventeenth century, even in the Dutch Republic, ruin meant not just destitution, but either consignment to the workhouse or starvation and possibly an early death. The last thing any of these people wanted to do was bid for another tulip. Every florist was a seller now.

That is not to say that prices fell instantly and simultaneously throughout the United Provinces. Some florists did move about from town to town, but most did not, so news took a day or two to travel. And in any case, the Dutch bulb trade was really a number of separate markets, one in each of the towns affected by the tulip mania. Prices in one city lagged behind those in another; the florists traded different bulbs; the tulip traders who met in one tavern were subtly different to those who made up every other college in the Republic.

Thus, while the tulip trade was in ruin in Haarlem, it continued to flourish briefly elsewhere. In Amsterdam – where news of the disaster in Haarlem must have reached the colleges by Wednesday – it was still strong on Friday 6 February, when a pound of Switsers was sold for 1065 guilders in the college of a tavern called the Mennonite Wedding. But the trade in Amsterdam seems to have approached a similar crisis point the next day, 7 February, when a florist named Joost van Cuyck bid 1100 guilders for another pound of the ubiquitous Switsers owned by Andries de Bosscher. Van Cuyck seems to have had second thoughts about the wisdom of this purchase, since he asked de Bosscher to guarantee him that the price would not fall. De Bosscher produced a colleague called Pieter van de Cruys, who was prepared to promise him 1200 guilders for the bulbs, but even this did not entirely satisfy van Cuyck. He seems to have doubted that van de Cruys had the means to make good his guarantee, and so on 11 February he went with de Bosscher to a local notary to put the whole agreement down in writing and make it legally binding. That must mean the deal was still alive eight days after Haarlem's florists had been unable to sell pound-goods at 1000 guilders a pound, and that the trade in Amsterdam survived for at least a week after the initial crash. Nevertheless, van Cuyck's evident concern suggests that trepidation quickly began to undermine confidence in the surviving centres of the tulip trade once the news from Haarlem had sunk in.

The same thing happened to the south, where lucrative trades were still being made in The Hague on 4 February. One that we know about involved Jan van Goyen, a well-known artist who was the most influential painter of landscapes in the whole of the United Provinces. Van Goyen was the son of a shoemaker, and his success as an artist had brought him a prosperity he could hardly have imagined in his youth. His father, a keen amateur artist who was at least affluent enough to own his own house, had suffered from fits of insanity and eventually had to be confined to the asylum at Leiden, leaving Jan to put himself through an apprenticeship with the Haarlem master Esaias van de Velde and make a name for himself with his paintings of sand dunes and river scenes. Although never really wealthy, he used the money that he earned to speculate in property and then in tulips, buying ten bulbs from Albert van Ravensteyn, a burgomaster of The Hague, on 27 January 1637 and then forty more eight days later for the combined sum of 912 guilders and two of his own paintings. Van Goyen's second purchase, agreed the day after the crash in Haarlem, was by far the bigger, being worth a total of 858 guilders. But shortly after the painter had agreed to buy the flowers, the market in The Hague slumped too, and he soon found himself in desperate financial difficulty.

The desperation of the bulb dealers was exacerbated by

the fact that the great majority of florists had participated, like Gaergoedt, in the *windhandel*, and when the market had collapsed they still remained liable to fulfil the futures contracts they had agreed. Practically every trader was now faced with a situation where he had put down deposits on tulips which were now worthless, and would be expected to pay substantial additional sums to complete his purchases when the bulbs were lifted in only a few months' time. Many had little choice but to default.

The collapse of the tulip trade thus had serious implications even for those who had sold their bulbs before the crash and appeared to have walked away with good profits. Among those affected in this way were the orphans of Wouter Winkel, who found themselves embroiled in at least two legal actions as a result of the Alkmaar auction. One involved a local dealer named Gerrit Amsterdam, who attempted to claim that the Verbeterde – the word means 'improved' – Boterman bulb of 563 aces he had purchased for 263 guilders had turned out to be nothing but a common Boterman, worth far less than he had paid. The other concerned Willem Lourisz., a florist from Heemskerk, just outside Haarlem. He had bid 512 guilders for a bulb of a *Rosen* variety called Anvers Vestus, the money to be paid when the tulip had flowered satisfactorily. A year and a half after the auction, Lourisz. still had not discharged his debt, and Jacob van der Meer and Jacob van der Gheest, the two regents of the Alkmaar orphanage who acted as guardians of the Winkel children, were eventually forced to sue him for nonpayment. The regents swore that they had invited the florist repeatedly to inspect his flower and settle his debt. Lourisz.'s defence – and it sounds a flimsy one – was that he had made an appointment to meet van der Meer outside the garden where the bulb was growing one morning in May 1637, that the regent had failed to show up, and that after waiting half an hour the florist had left. Van der Meer indignantly countered that there never had been an appointment, that the tulip had flowered magnificently and had been available for inspection for several weeks, and that it should be paid for as agreed.

The position of the growers was a little better than that of the florists. Even after the tavern trade had come to a halt, connoisseurs did continue to pay extraordinary prices for tulips. On 17 March a Haarlem merchant named Dirck Boortens sold a quantity of bulbs, including an Admirael Liefkens and a Saeyblom, to one Pieter van Welsen for 11,700 guilders. Van Welsen went to inspect his flowers in the middle of April and discovered some of them were in poor condition, so Boortens agreed to knock 300 guilders off the original price. Van Welsen cannot have been too worried about the collapse of the tulip colleges, for he confirmed that he was still perfectly content to pay the balance of 11,400 guilders and agreed to do so in three instalments: 4000 guilders in June, 3700 at the end of August, and the remaining 3700 guilders on the first day of February 1638. This was a private agreement between two genuine flower lovers who had probably never bothered themselves with the tulip mania going on around them and who could well afford the fantastic luxury of spending 10,000 guilders or more on plants that would be enjoyed for only a few weeks each year. But even among the more affluent and high-class tulip traders there were still occasional signs of optimism about the state of the market. Jan Quaeckel, the rich Haarlem grower who owned the Golden Grape, travelled to the auction at Alkmaar the day after prices crashed in the colleges of his home town and still felt confident enough to part with 3260 guilders for some of Wouter Winkel's fine bulbs. And in May 1637, Jan Admirael – the fashionable Amsterdam dealer who grew tulips in the garden behind his house on the exclusive Prinsengracht – agreed to guarantee his customer Paulus de Hooge that he would make at least 20 per cent over the next twelve months on any tulips he bought from Admirael's flower beds.

Yet the collapse of the tavern colleges threatened to bankrupt not just the florists but a substantial number of the bulb farmers too. Any grower who had been tempted, during the mania, to expand their business and sell to florists as well as connoisseurs was affected, and the crisis was serious enough to prompt the professionals to unprecedentedly rapid action. As early as 7 February, only four days after the crash in Haarlem, the growers of the provinces of Holland and Utrecht agreed to arrange a general meeting, to be held at Amsterdam, to discuss ways of minimizing the damage caused by the collapse in prices. Even in a country the size of the United Provinces some of these towns were two days' journey from each other, so the growers' reaction was astonishingly rapid. It can only have been the result of the most acute concern about their future.

With the single exception of Rotterdam – whose growers sent a letter agreeing to be bound by the decisions of the majority – each of the dozen towns and districts most closely involved in the tulip trade held their own assemblies to elect representatives to attend this meeting. Most of the big bulb growers, including Francisco da Costa from Vianen,

Barent Cardoes and Willem Schonaeus from Haarlem, and François Sweerts from Utrecht, travelled to Amsterdam alongside lesser-known names such as W. J. Sloting of Leiden and Claes Heertgens, who was one of the representatives from the Streeck, a strip of good bulb-growing land which lay between the three West Friesland towns of Hoorn, Enkhuizen and Medemblik.

The great assembly of growers took place on 23 February. By then the tulip trade must have been everywhere in utter disarray, because the delegates seem to have wasted little time considering whether there was any way it could be revived. Instead, their discussion centred on ways of cutting their losses.

In some respects, the growers' problems were nearly as bad as those faced by the florists. The great majority of them had incurred considerable costs over the previous year in buying bulbs and offsets, cultivating their gardens and, quite possibly, expanding their operations to try to meet the sharp rise in demand. Now they were faced with a situation where they were owed huge sums of money by customers who had paid only small deposits and had long since sold the tulips on to other traders. In many cases, the rights of ownership had vanished into one of the long and complex chains of trades and deals created during the mania. If just one of the florists involved in one of these agreements found himself unable to settle his debts, the whole chain would collapse and the growers waiting at one end of it would have not the slightest prospect of receiving the balance owing on their bulbs when payment fell due in June.

All these problems must have been debated during the

assembly in Amsterdam. The growers' solution was to pretend, in essence, that the mania had never happened. As the assembly drew to a close, the majority backed a resolution which held that transactions made before the last planting should still be binding, and that while buyers could have the right to cancel any purchase made since 30 November 1636, they should be required to pay 10 per cent of the sale price by way of compensation. The representatives of Amsterdam were the only ones who refused to sign the agreement.

In promoting this compromise, the growers were rather cynically attempting to minimize the losses that they faced. The majority of the bulbs that were sold before the end of November, they knew, had been bought by connoisseurs and wealthy dealers who had the means to pay their debts in full. It was only in December and January that poorer florists had flooded into the market as the tulip trade exploded and full-fledged mania took hold. Obtaining payment from these people would be a very different matter, and the resolution agreed at Amsterdam recognized that fact.

In the *Samenspraecken*, Waermondt explains to Gaergoedt how the growers' plan would work in practice. If a bulb sold originally for 30 guilders had been resold three times, say for 60, 100 and 200 guilders, then the man who had offered 200 guilders had the choice of handing over the money and keeping the flower. If he no longer wanted it, he would have to pay 20 guilders to the man who had sold it to him in order to have their agreement annulled. The right of ownership would then revert to the florist who had bid 100 guilders, and he in turn would have to choose whether to keep the bulb for himself or pay compensation of 10 guilders to the man from whom he had purchased the tulip. Waermondt does not say so, but presumably the growers' intention was that if any one contract was settled in full, all those beneath it in the chain – including the original sale agreed by the grower – would also be honoured. If none of the florists wanted to keep the bulb, possession would revert to the grower, who would receive 10 per cent of the sale price as compensation. He then had every right to sell the tulip to another buyer if he could.

The reasons why the growers of Amsterdam declined to ratify this agreement are not known, but it is conceivable that they were shocked by the sheer scale of the surrender their fellow growers were proposing. After all, the bulb farmers had every right, in the eyes of the law, to pursue claims to full payment for their produce. Only sheer pragmatism, an acceptance that – whatever the rights and wrongs of the matter – it would be a waste of time to pursue hundreds of insolvent debtors through the courts, can have persuaded the majority who agreed the growers' resolution to voluntarily sign away their right to what must, in many cases, have been several thousands of guilders. Ten per cent – by which they meant one-tenth of the original sale price, not the final value of the tulips when the mania reached its peak – must have been all most bulb farmers believed they had a chance of recovering from the disaster.

The growers' problem was that even this modest demand for part payment could hardly be enforced. They could ask their customers to consider the compromise, but they could not require them to accept it. And since most florists could not hope to find money enough to cover even one-tenth of their liabilities unless they were paid for the bulbs they in turn had sold to others, there was little prospect of many of them coming to terms with the growers unless they had to. 'When my buyer pays me, I will pay you,' Gaergoedt assures a creditor in the *Samenspraecken*. Then he adds an ominous caveat: 'But he is nowhere to be found.'

It was thus evident that the bulb trade's troubles could not be resolved by the bulb trade alone. Some higher authority would have to rule on who owned the thousands of tulips bought and sold before February 1637 – and, more importantly, who should pay for them. Whatever compromise was finally proposed, moreover, would need to have the force of law.

The mania had become a problem for the courts. But while the tulip's case was being heard, the critics of the flower were going to have their say.

CHAPTER 13

Goddess of Whores

No one in the United Provinces loved tulips more than Claes Pietersz. of Amsterdam, who was probably the most fashionable physician in the whole of the Republic. Other men might grow the flower, trade it and even make their fortunes from it. Pietersz. changed his name because of it. He became, quite literally, Dr Tulip.

Claes Pietersz. began styling himself Nicolaes Tulp (the Dutch word for the flower) in 1621, when tulips were just coming into vogue among the wealthiest and most discerning members of the regent class. He used the flower as a personal emblem, too. When he was elected an alderman of Amsterdam in 1622, and had to choose a coat of arms, Tulp had his shield adorned with a delicate, scarlet-flamed *Rosen* tulip. His alderman's seal stamped a red wax flower on the hundreds of official documents to which he gave his approval. And when he returned home after a long day in the service of the city, it was to a painting of the tulip – one of the finest of the fabulous Admiraels, it was said – which adorned a signboard swinging to and fro over the door of his fashionable house on the Prinsengracht.

In time, young Dr Tulp (he was in his late twenties when he changed his name) rose to a position of great eminence. He became a friend of Rembrandt and was the subject of one of the painter's most celebrated canvases, *The Anatomy Lesson of Dr Tulp*, which shows him as a distinguished physician – a tall, slender man, all angles and beard – busily dissecting the body of a recently executed criminal. Contemporaries knew Tulp as a botanist, a vigorous promoter of the medicinal benefits of tea – which he prescribed as an antidote to lassitude and cramps – and a successful politician who was four times burgomaster of Amsterdam. He was also a notoriously stern Calvinist, whose principled disdain for the intoxicated revels which were traditional at even the grandest of Dutch weddings led him to sponsor a piece of legislation for which he is still occasionally remembered today: Amsterdam's 'sumptuary law' of 1655, which made it an offence for wedding feasts to involve more than fifty guests or last longer than two days.

It is hardly surprising, then, that Dr Tulp instinctively hated the drunken excesses of the tavern colleges. In private he remained a connoisseur until the end of his long life; indeed in 1652, on the occasion of his retirement from the surgeons' guild, he presented a silver beaker in the shape of a tulip, with a lizard climbing its stem, to his old colleagues and asked that in future it be used to propose the final toast at the guild's numberless banquets. But in public, after 1637, Nicolaes Tulp preferred not to be associated with the famous flower with which he shared his name. The sign outside the house in the Prinsengracht came down, the coat of arms was less prominently displayed. Dr Tulip felt ashamed of the excesses of the tulip mania.

There were many who shared Tulp's sentiments. Adolphus Vorstius, the professor who occupied Clusius's old chair of botany at the University of Leiden and lectured twice each week in the *hortus* there on the properties of its plants and herbs, came to despise the vulgarity of the traders and their hysteria for bulbs, and took to destroying every tulip he came across, hacking away at the flowers with a staff. Even outsiders who had taken no part in the mania themselves often shared the connoisseurs' low opinions of the florists. During the final stages of the bulb craze, many ordinary people began to refer to the members of the tavern colleges, derisively, as '*kappists*'. This was a considerable affront; for the Dutchmen of the Golden Age, the name summoned up the image of a fool clad in a jester's cap.

Not all the critics of the tulip mania confined themselves to jokes and insults. Some, particularly the more religious members of Dutch society, took an altogether sterner view, accusing bulb dealers of casting aside the Christian principles of charity and moderation. And even before the final collapse of the market for tulips, a number of the most vociferous opponents of the mania had gone into print with their criticisms of the bulb trade. Their medium was the pamphlet, and beginning in the final months of 1636, presses throughout Holland poured forth a flood of broadsides on the subject of the flower craze.

Most of these productions were ribald satires. With few exceptions their central character was the Roman goddess Flora, who had always been the most licentious of deities. According to her myth, Flora had been a notorious courtesan in the earliest days of Rome who left so much of her immoral earnings to the city when she died that the grateful Romans deified her. She became both the goddess of flowers and the protector of prostitutes, and Dutch pamphleteers delighted in nothing more than drawing obvious parallels between the Roman whore and the valuable tulips which had been passed from hand to hand so rapidly at the very height of the mania. Flora, they reminded their readers, had made a practice of selling herself to the highest bidder, and her price had risen constantly until it was so steep no man could afford to keep her to himself for long. Though each of her lovers was richer and more generous than the last, she ruined them all with her demands for ever more lavish proofs of their devotion. Even after she ascended to the Latin pantheon and married the west wind, Zephyr, Flora had proved incapable of mending her ways. Before long she cuckolded her new husband by dallying with Hercules.

Faithless companion, grasping mistress: perfect metaphor. In the eyes of the pamphleteers, Holland's bulb traders were just the latest in a long line of men who had abandoned themselves to the goddess of whores, only to be betrayed by her. Many of their publications alluded to the florists' dire financial straits and bore titles such as *Flora's Sick-Bed* or the somewhat more explicit *The Fall of the Great Garden-Whore, the Villain-Goddess Flora*. Others contained the fictional complaints of traders who had found themselves in thrall to a false and pagan idol. In one broadside, a weaver speaks angrily of how Flora seduced him. In another, revealingly titled *Charge Against the Pagan and Turkish Tulip-Bulbs*, Flora and the other earth spirits decree that the tulip and all other herbs and plants should return to their original places in the scheme of creation, on pain of plagues of vermin and foul weather let loose on the land. The overall tone is one of bitter antagonism towards a goddess who had promised everything, yet left those foolish enough to trust her with less than nothing.

At the same time that pamphleteers were pouring forth a torrent of sarcastic verse, the first of several memorable works of art, each rich in the details of the tulip trade, appeared, revealing more about the sort of ridicule the ruined florists must have endured after the crash. It was a painting by Pieter Nolpe (which was later turned into a copper engraving by an artist named Cornelis Danckerts) ponderously entitled *Flora's Fool's Cap, or Scenes from the Remarkable Year 1637 when one Fool hatched another, the Idle Rich lost their wealth and the Wise lost their senses*. Nolpe's work shows bulb dealers gathering in a drinking house called At the Sign of the Foolish Bulbs, which is actually a gigantic jester's cap. The sign outside the inn shows two men fighting. In the foreground, men carrying baskets and pushing wheelbarrows full of bulbs are on their way to dump the now worthless tulips upon a dung heap; three gardeners stand and watch, while just behind them Beelzebub, armed with a fishing rod, casts about for worthless tulip contracts. In his right hand, the devil holds an hourglass which indicates that the sands of time have quite run out for the tulip trade. In the background of the picture stands a derelict house, and the goddess Flora can be seen riding past on a donkey, gesturing at the members of an angry crowd to keep their distance. She is, the text below the picture explains, being driven off 'for her whorish immorality'.

Similarly pointed attacks on the excesses of the bulb trade

continued to appear for years afterwards, so this artistic evidence supports the contention that the mania had a considerable impact even on those who had taken no active part in it. In 1640 Chrispijn van de Passe (the same van de Passe whose *Hortus Floridus* had helped to establish the fashion for tulips more than twenty years earlier) engraved a famous illustration entitled *Floraes Mallewagen* ('Flora's chariot of fools'). This picture shows the goddess, drawn as a young girl in blooming health and a low-cut dress, riding a luxuriously appointed sand-yacht packed with carousing *kappists* dressed in jester's caps. These allegorical figures bear labels such as 'Vain hope', 'Tippler' and 'Hoard it all'. The sand-yacht itself is drawn tearing across the beach outside Haarlem, and is adorned with the signs which hung outside some of the local taverns involved in the mania – the White Doublet, the Little Hen and four or five others. An ape climbs the mast and defecates over the florists below. Flora, who is seated in state in the stern of the vessel, carries a bunch of the most sought-after tulips: Generael Bol, Admirael van Hoorn and (of course) Semper Augustus in one hand; others, including a Gouda and a precious Viceroy, wait on the sand to be crushed beneath the sand-yacht's wheels. The contraption is heading straight for the sea, but a crowd of would-be tulip dealers run behind the yacht, desperate to join it on its short rush to destruction. They are weavers and, in their haste, they are trampling underfoot all the tools of their old profession. In the four corners of his engraving, van de Passe placed small insets. One shows the bulb grower Henrik Pottebacker's famous garden at Gouda, the others tavern trading scenes in Haarlem and Hoorn. The central feature of his piece, the fast-moving sand-yacht, is itself a powerful metaphor for the fatal wind-trade.

In the same year that van de Passe engraved his fools'-chariot, Jan Breughel the Younger painted an ambitious work entitled *Allegory Upon the Tulip Mania*. Breughel was the most influential painter of flowers to emerge during the Golden Age. Although some modern critics find his style a little stiff, his flower paintings are always vivid and enlivened by the inclusion of small details, such as insects crawling upon the leaves. The *Allegory* is an exceptionally lively piece, as packed with incident as any cartoon by Cruikshank or Gillray. Two dozen simian florists are portrayed indulging in all the rituals of the bulb trade. One points at some flowering tulips; another holds up a flower in one paw and a bag of money in the other. Behind them a group of monkeys fight over who should pay for the now worthless bulbs and one speculator is carried to an early grave. On the right hand side of the picture, a pair of apes share one of the florists' traditional banquets while another is hauled before a magistrate for defaulting on his debts. In one corner, a particularly disgruntled monkey urinates on a flower bed full of tulip bulbs.

These scabrous satires undoubtedly had a huge impact. Even one hundred years later, the tulip mania remained a raw and vivid scar upon the national psyche of the Dutch and, thanks in good measure to the pamphleteers and painters of the Golden Age, the very idea that bulbs could ever have been traded for colossal sums strikes many as perfectly ridiculous today. Nevertheless, the pamphlets of the mania, at least, are important not so much for what they are – ephemeral single sheets, often enough, which were illustrated with one shoddy woodcut, quickly and cheaply printed on low-quality pulp, and peddled by hawkers for a few stuivers apiece – as for the reasons they were produced. A few had been written simply to entertain; in the Dutch Republic, where literacy rates were high, pamphlets were a useful and profitable sideline for men like Adriaen Roman, the official government printer of Haarlem. Roman, who published the three dialogues between Waermondt and Gaergoedt, could hope to sell perhaps 1000 or 1250 copies of a typical broadside, and best-sellers such as the *Samenspraecken*, which were reprinted on several occasions, could reach as many as 15,000 people.

The majority, though, were produced specifically to influence public opinion. Pamphlets of the latter sort were typically funded by wealthy men who lacked the literary skills to pen something of their own. Instead they paid hack writers to put their views into verse and printers to publish and distribute the results. The actual authors of these works – men such as Stephen van der Lust, a professional playwright from Haarlem who churned out four pamphlets on the mania, and Jan Soet, a satirist with a vicious pen who wrote two – were often impoverished writers who wrote in rhyme or dialogue in order to appeal to the common man and intended their words to be read aloud to audiences gathered in taverns and other meeting places. Their shadowy patrons, on the other hand, were generally regents and patricians who had their own very specific agendas.

Seen in this way, it becomes clear that a number of the pamphlets published in the Dutch Republic in the spring of 1637 were religiously inspired works. The patrons who caused these broadsides to be published saw the bulb trade as irreligious and immoral, and condemned those who had participated in the mania for indulging a sinful appetite for profit. A smaller number of pamphlets, on the other hand, seem to have been designed to drum up support for the old growers and connoisseurs, who had been just as horrified by the mania as the sternest critics of the bulb craze. These broadsides, which bore tellingly defensive titles such as *A new song about the connoisseurs who don't go to the tavern and because of that wish to be distinguished from the florists*, attempted to show that true tulip lovers bore no responsibility for the mania and were still deserving of respect. On the whole, though, their arguments must have sounded hollow to those who looked on the whole bulb trade with horror and distaste, and it is difficult to believe that it was not the harder-hitting and more vitriolic broadsides that were the better sellers.

While the writers and artists of the United Provinces poured scorn on those who had lost everything they owned to tulip mania, the authorities of the Republic were coming slowly to terms with the problem of averting financial catastrophe.

The first difficulty was deciding who should resolve the thousands of outstanding tulip contracts. The only certainty was that the vast majority of these agreements would have to be nullified; in almost every case, the would-be buyers no longer had the desire or, more importantly, the money to fulfil them. But whether the bulb contracts should be cancelled on the terms proposed by the growers – 10 per cent of the agreed selling price – or those favoured by the florists (who hoped to pay nothing) was another matter altogether.

In normal circumstances it would have fallen to the regents of each of the towns caught up in the mania to decide which proposal to accept, or to substitute a solution of their own. But so far as the governors of these cities were concerned, the mania had the makings of a particularly tricky problem, and their response was far from resolute.

In Haarlem, the town we know most about, the city council approved three separate resolutions in the space of little more than a month, proposing that disputes between florists be resolved in three different ways. The regents' first decree, issued on 7 March, annulled every transaction which had taken place within the jurisdiction of the city since the previous October without, apparently, making provision for the payment of any sort of compensation to the sellers. Less than five weeks later, in a second resolution which effectively reversed the first, the city fathers ruled instead that 'those persons who have bought tulips in eating-houses will be obliged to pay for their transactions'. (The councillors did not explain how thousands of nominally bankrupt florists would find the money to comply.) Then, within a week of publishing that decree, Haarlem's regents changed their minds for a third time. On this occasion, instead of proposing yet another solution, they resolved to wash their hands of the matter. They referred the whole problem to their immediate superiors, the members of the provincial parliament, the States of Holland, sitting at The Hague, petitioning the States for a ruling and suggesting that it adopt the compromise originally suggested by the growers at their meeting of 23 February.

Such indecision was quite uncharacteristic of the sober governors of Haarlem, and in all likelihood the changes in the city's policy were the product of vociferous lobbying by the various interested parties: growers demanding the right to seek full payment, florists begging for relief. The subject must have been endlessly debated throughout the spring of 1637, and the councillors repeatedly harangued by tulip dealers anxious to press their own solutions to the problem. The frustration that they felt is illustrated by a resolution of 17 March which banned both the printing and the sale of inflammatory pamphlets on the mania and ordered the booksellers and printers of the city to surrender their stocks of the offending broadsides to be burned. The regents' willingness to hand the matter over to a higher authority suggests they recognized the impossibility of conjuring a compromise acceptable to all.

Similar protests probably occurred elsewhere, and other Dutch cities joined Haarlem in petitioning the States of Holland to find a solution which minimized the losses to both growers and florists. By the middle of March, the burgomasters of Hoorn were asking their representatives in The Hague to do what they could to speed the decision-making process. But the States, like the cities, soon realized that the tulip mania was a novel problem, and one which required careful consideration. Its members had little information on which to base a solution; judging from the example of Haarlem, where only two of the fifty-four regents who governed the city in 1636–7 had any involvement in the mania, few if any would have participated in the bulb trade themselves, and the scant summaries of events which some cities appear to have forwarded to The Hague cannot have provided sufficient detail. The States preferred to refer the matter to the province's supreme judicial authority, the Court of Holland. It called for further information, and, while it waited, it turned its attention back to other matters.

For more than a month, then, from the middle of March to the end of April, everyone caught up in the mania – growers and florists – endured an agony of anticipation. Tulips which had been worth fortunes only a few weeks earlier were in flower throughout the United Provinces, but while they bloomed, brightening the damp Dutch spring, hundreds of florists were consumed with the fear that they would be forced into bankruptcy, and thousands of agreements, worth millions of guilders, remained unsettled.

For those who had actually participated in the mania, the most pressing concern was to survive the impending financial catastrophe. But they also wanted to understand why the market had crashed. Of course, few admitted, even to themselves, that they bore responsibility for their plight. They preferred to see themselves as victims, and, like victims everywhere, they found explanations that exonerated them from blame.

Many came to believe that the bulb craze had been some sort of fraud. At one extreme were those who simply thought they had been cheated by their fellow florists, or perhaps by the auctioneer at their college. At the other stood a smaller group of people who had convinced themselves that the tulip trade was itself a conspiracy. One anonymous author suggested that the market had been created and controlled by a shadowy cabal of twenty or thirty of the richest growers and dealers, who had deliberately manipulated prices to their own advantage. How such a group could possibly have hoped to coordinate their activities across the dozen towns infected by the mania was not explained.

Blame for the mania was also placed elsewhere. The same

writer who had hinted at the existence of a cabal also suggested that some of the worst excesses of the tulip trade were the result of the manipulations of bankrupts, Jews and Mennonites, three groups that stood apart from the rest of society and thus made convenient scapegoats. Bankrupts, after all, had failed to adhere to the sacred Dutch principle of living within one's means, had been forced to account for their transgressions, and might well be looking for revenge. Jews, though considerably better treated in the United Provinces than in Germany or France, were nevertheless closely connected in the popular imagination with money-lending and other forms of profiteering, and they had long been prohibited from mixing too freely with the rest of the population; the men were actively discouraged from conversing with Dutch women, and it was illegal for them to hire Christian servants. Mennonites, too, were outsiders. They were an Anabaptist sect easily distinguishable by their dress (they clothed themselves entirely in black, favouring long jackets and baggy breeches). As well as opposing infant baptism – something the orthodox Dutch regarded as both a moral obligation and a necessity at a time when child mortality was still extremely high – Mennonites were pacifists who steadfastly refused to bear arms. This made them unpopular at a time when the United Provinces were still at war with Spain.

None of these accusations stands up to scrutiny; indeed, there is no real evidence that any group – other than, perhaps, the tulip growers themselves – had promoted the bulb craze to further their own ends. It is true that some Mennonites had involved themselves in the mania; one, Jacques de Clerq, a merchant who traded with the Baltic

and Brazil, was buying and selling tulips for as much as 400
guilders a bulb as early as the winter of 1635. But many
other members of the sect were highly critical of the tulip
trade, and urged those who dealt in bulbs to cease. Similarly,
there were actually very few Jews in the United Provinces,
and the only one known for certain to have been heavily
involved in the tulip trade – the renowned Portuguese
grower Francisco da Costa – appears to have been a man
of unblemished reputation. As for bankrupts, none of the
records of the time suggest that a single one played any part
in the mania.

Probably only a minority of florists believed in such
conspiracy theories. But a number do appear to have
suspected that individual dealers had forced prices up arti-
ficially in order to maximize their profits. Price-fixing was
popularly supposed to be accomplished by the age-old
means of fake auctions. These affairs were supposedly
organized by cunning traders who opened the proceedings
by 'selling' bulbs for record prices to their own accomplices
in order to stimulate excitement and persuade others to buy
at inflated rates.

A number of florists laid the blame for the mania at the
feet of the growers. Some were accused of stoking up
interest in tulips by selling bulbs with a guarantee that they
would buy them back next year for more than they had
cost. Others, it was claimed, passed off worthless *vodderij* as
valuable bulbs. A grower from Amsterdam who was suspected of this sort of fraud is said to have tampered with
the bulbs he sold, running them through with needles in
order to damage them so badly they would not flower and
reveal his deception. The man was eventually caught when

one disgruntled purchaser made a close inspection of his tulips and discovered tiny puncture marks on the surface of the bulb.

It is perfectly possible that methods of this sort were indeed practised upon occasion, but surely not so cynically and so regularly as to have a significant effect on bulb prices. In truth, there was no need to concoct elaborate conspiracy theories to account for the excesses of the bulb craze. Given the fundamental discrepancy between demand for bulbs and the limited supply, greed, inexperience and the short-sightedness of the florists themselves were all that was required to turn tulip trading into tulip mania.

It was the last week of April before the Court of Holland finally concluded its review of the tulip mania. Eight weeks had passed since the growers had met at Amsterdam to propose their own solution to the crisis, three months since the collapse of the flower trade throughout the province. Yet when the learned judges of the Court returned their findings to the States, they began by admitting that they still did not fully understand what had caused the bulb craze or why things had got so badly out of hand.

The Court of Holland was, however, certain of one thing: it wanted as little as possible to do with the tangled and intractable wrangles thrown up by the mania. Instead, it recommended that disputes between buyers and sellers, florists and growers, should be referred back to the towns to be dealt with locally wherever possible. The Court suggested that city magistrates should begin by gathering detailed information about the flower trade. Only when they had a better understanding of what had happened in their towns should they hear disputes, and while the necessary data was collected, all contracts for the purchase of bulbs should be temporarily suspended. If, in the event, there were cases which could not be dealt with at a local level, they might still be referred to The Hague; but this, it was implied, was a remote contingency. The Court's verdict was clear: the cities should solve their problems on their own.

Presented at last with some definite suggestions, the States of Holland wasted little time in acting on them. On 27 April, only two days after the Court presented its proposals, the representatives at The Hague agreed a resolution which incorporated all the main recommendations and made them binding on the cities of the province. A letter explaining the resolution was sent by fast messenger to all the towns of Holland. Thus, by 29 April, the burgomasters of each of the cities affected by the mania finally received instructions on how to deal with the hundreds of disputes still awaiting resolution.

The key point was the Court of Holland's suggestion that all contracts for the sale of bulbs be suspended while the mania was thoroughly investigated. As originally proposed, this recommendation was plainly intended as a temporary measure; indeed the Court acknowledged that, once they were properly informed, local magistrates might decide that the contracts signed in the colleges could be enforced. In that event, it noted, disgruntled sellers should be permitted to pursue defaulting customers for payment. Though time went by, the towns involved in the bulb craze never did compile detailed information on the tulip craze as the Court requested, and no further action was ever taken at The Hague. The regents of the States of Holland were more than happy to wash their hands of the entire affair. What had been intended as an interim measure became the basis for the liquidation of the mania.

This was very good news for the florists. Most cities implemented the States of Holland's resolution by ordering their solicitors and magistrates to have nothing more to do with the mania. In Haarlem, for example, the regents who governed the town instructed that the solicitors and notaries should cease to issue writs on behalf of tulip traders, and the messengers who normally served protests and summonses were instructed not to handle any that related to the bulb craze. Similar orders were issued in Gouda and the three West Frisian towns of Enkhuizen, Medemblik and Hoorn.

The florists of these cities who believed they had no option but to default on their obligations could now do so without fear of retribution, and hundreds of poor artisans who had more than half expected they would be forced into bankruptcy took full advantage of this fantastic piece of good fortune. A handful of those caught up in the mania were rich and honourable enough to meet their obligations, it is true, including the Alkmaar man who had bought 7000 guilders-worth of bulbs from Henricus Munting and now exercised his right to pay just 700 guilders to cancel the contract and return the tulips to their original owner. But, as the Haarlem solicitor Adriaen van Bosvelt cynically observed, honest florists were hard to find. Throughout Holland, van Bosvelt wrote, 'a great number of persons [are] unwilling to pay or come to a compromise'. Even those who did offer to settle at least part of their debts did not come close to parting with the 10 per cent the growers wanted. The handful who did pay a little offered no more than 'one, two, three, four, yes, even five, which was the utmost, out of a hundred'.

The blanket ban on tulip cases quickly had the desired effect. Growers and florists were forced to settle their disagreements among themselves, and the regents ceased to be bothered by the fallout from the mania. But even now, it was a long time before the last dispute was settled. We know that in Haarlem the process of liquidation dragged on through 1637 and for the whole of 1638, not least because some tulip traders proved more reluctant to settle their differences than the States of Holland hoped. It was probably the same in other cities.

In the event, many of those caught up in the mania did seek their own solutions, as the regents had hoped. A large number of agreements were cancelled with the consent, if not the approval, of all the parties concerned; in Alkmaar, in fact, all tulip contracts appear to have been nullified in this way. The growers did what they could to recover their losses by putting thousands of bulbs that had never been collected up for sale. (Not surprisingly, few people were still interested in buying them, but a handful of the rarer tulips did eventually sell to connoisseurs for decent sums.) And the unfortunate Haarlem dyer Jacob de Block, who had been required to honour his guarantee to Geertruyt Schoudt, took his pound of unsaleable Switsers off to Amsterdam in the hope of disposing of the bulbs there.

Some, though, were determined to fight for their lost fortunes. The most fortunate were those who had bought and sold bulbs in the colleges of Amsterdam, which – apparently alone among the towns caught up in the mania –

still allowed tulip cases to be brought before its courts, and within a few weeks a few of the growers of the city began to take advantage of this dispensation to sue their former customers.

One of the most active litigants was Abraham de Goyer, the scion of an old regent family and a grower who kept at least two gardens: one on the Cingel just outside Amsterdam's Regulierspoort and the other on the Walepadt, by the city walls. On 10 June he demanded 950 guilders from one Abraham Wachtendonck for the four bulbs of Late Bleyenburch and the pound of Oudenaers that Wachtendonck had purchased the previous autumn. The next day de Goyer began an action against Liebert van Acxel, who had agreed on 1 October to buy the offsets of a De Beste Juri and a Bruyn Purper for 1100 guilders and a Purper en Wit van Quaeckel (one of old Cornelis Quaeckel's *Violetten* creations) for 750 guilders. In order to bolster his case, the grower asked a notary named B. J. Verbeeck to accompany him to his garden on the Walepadt, where the two men lifted all the bulbs and confirmed that the Purper en Wit van Quaeckel and the Bruyn Purper had developed two offsets apiece. De Goyer seems to have expected trouble with yet another of his other customers, since he also asked Verbeeck to confirm that he had lifted an Admirael Liefkens with one offset which he had been growing in the garden of a man named Willem Willemsz.

A few other growers with business in Amsterdam also took the opportunity to assert their rights. Hans Baert of Haarlem sought 140 guilders for the 2000 aces of Groote Gepumaseerde he had sold to Hendrick van Bergom of Amsterdam. Jan Admirael, who had gone to such lengths to persuade Paulus de Hooge to buy his bulbs, changed his tune when de Hooge failed to pay the money he owed and sought the advice of his solicitor. And Willem Schonaeus, of Haarlem, demanded nearly 6000 guilders from François Koster, the balance of the sum the hapless Koster owed on a substantial quantity of *vodderij* and a handful of piece-goods he had ordered on 3 February:

Four pounds of Switsers	6000 guilders
2000 aces of Maxen	400 guilders
1000 aces of Porsmaeckers	250 guilders
	6650 guilders

Nor did every florist accept the ban on tulip cases in cities such as Haarlem. A handful found pretexts for bringing their disputes to court in different guises. One such case got under way in November 1637: having waited until the last possible moment before his bulbs had to be replanted, in the vain hope of receiving his money, a local grower named Pieter Caluwaert knocked on the door of the merchant Jacques de Clerq and attempted to hand over the pound of Witte Croonen, two pounds of Switsers, five Oudenaers and three Maxen that he had agreed to purchase nearly a year earlier. When de Clerq did his best to avoid his former colleague, Caluwaert began proceedings against him, presumably on the grounds that he had refused to accept a delivery.

All in all, though, only a tiny minority of tulip cases ever found their way to court, even in Amsterdam. The reason was simple: few of the florists possessed enough money to be worth suing. De Goyer, Admirael and Baert sought payment from wealthy customers who possessed the means to pay their bills. The great majority of the florists who had been caught up in the mania were not so well off, and there would have been little point in dragging these people through the courts.

Even so, there were still many who refused to tear up their tulip contracts and accept their losses. At the end of January 1638, a full year after the crash, hundreds of cases remained to be resolved. These disputes were proving highly disruptive; they soured relations between people who had once been colleagues or friends and were a constant and embarrassing reminder of the excesses of the mania, which was a nightmare most Dutchmen were anxious to forget. There seemed little prospect, moreover, that they would ever be resolved unless the local authorities took some further action.

On 30 January, therefore, the governors of Haarlem set up an arbitration committee to consider the remaining tulip cases. Panels of this sort already existed throughout the United Provinces; the arbitrators were commonly called 'friend-makers' and, as Sir William Brereton discovered on his tour through Holland in 1634, they could be found in most Dutch cities and were specially chosen for their integrity and common sense. The friend-makers, Brereton discovered, 'had authority to call any man before them that hath any suit or controversy; they are to mediate in a friendly manner, in a way of arbitration, and are to compose and conclude differences.' They had the additional advantage that, unlike the traditional courts, they offered their services for free.

Some of the records of a similar arbitration court which

was set up in Amsterdam have survived to indicate the sort
of verdicts that the friend-makers handed down. In one
case, contested between Jan Admirael and Wilhelmus Tyberius, the rector of the Latin School at Alkmaar, the arbitrators ordered Admirael to pay Tyberius 375 guilders to
settle their differences. The terms of the arbitration, though,
were fairly generous; the Amsterdam grower was given ten
months to come up with the money and mildly requested
to let that be an end to the matter.

At first the burgomasters of Haarlem granted their friend-
makers only limited powers to resolve the outstanding tulip
cases. The new panel, which had five members, sat at least
twice each week and could subpoena witnesses who were
reluctant to appear before it. But its decisions were not
binding, and many warring florists proved reluctant to accept
the compromises it recommended. From the surviving
evidence, it appears that little progress was made in working
through the backlog of disputes in Haarlem.

It was only in May 1638 that the regents of the city finally
took the matter properly in hand and issued – for the first
time since the abortive growers' meeting almost eighteen
months earlier – definite guidelines for resolving all outstanding disputes. Buyers who wished to free themselves of
their obligations, the city council ruled, could cancel their
contracts by paying 3.5 per cent of the original sale price.
Ownership of the bulbs would then revert to the growers.
This was the most affordable and workable compromise yet
suggested, and the council backed it up by ruling that the
friend-makers' verdicts should henceforth be binding in all
cases.

The compromise meant that even florists with debts

running to thousands of guilders could clear their obligations by paying 100 guilders or less, a sum that even the poorest could pay off in instalments. And, while inherently unfair to the growers, it did guarantee them a minimal payment which in all likelihood covered their costs and left them little worse off than they had been before the mania erupted.

The tulip mania thus ended, as the Court of Holland had wished, not in a flurry of expensive legal actions but in grudging compromise. In the end it had been a craze of the poor and the ambitious which – contrary to popular belief – had virtually no impact on the Dutch economy. No general recession followed in its wake, and the vast majority of florists emerged from the liquidation shaken and chastened, but little better or worse off than they had been before the mania began. Their paper profits and their paper losses effectively cancelled each other out, and even the richest florists were never formally punished for defaulting on their obligations.

Indeed, by far the most striking thing about the handful of cases that did find their way into the hands of the solicitors of the province was that there were no celebrated trials, no verdicts and no records of any convictions. The growers and their customers invariably settled their differences out of court. Even in Amsterdam, the liquidation of the mania was not a legal matter but a process of compromise and reconciliation agreed by the florists themselves.

The last known case resulting from the bulb craze was heard in Haarlem on 24 January 1639, almost exactly two years after the crash. A grower named Bruyn den Dubbelden demanded 2100 guilders from his customer Jan Korver, of Alkmaar, payment for a pound of Gheele Croonen at 800 guilders and two pounds of Switsers worth 1300. No verdict is recorded. Presumably this means that den Dubbelden, like other growers, was compelled to settle at 3.5 per cent, and that a contract that had been worth seven years' wages to a Haarlem artisan only a couple of years earlier was cancelled for a payment of 73 guilders, 10 stuivers.

Even now, a tiny minority of cases remained unsettled for reasons which have been lost to history. The hapless artist Jan van Goyen was one of the unfortunate few who continued to suffer for dabbling in the bulb trade. For the rest of his life, burgomaster van Ravensteyn pursued his former customer relentlessly for the whole of the money he was owed. Van Goyen handed over one of the pictures he had promised, but he had invested almost all his available capital in tulips and with the crash in prices he no longer had any prospect of repaying his debts. Having produced little art in the three years he had devoted to speculating in the property and bulb markets, the painter was forced to return to his easel.

The simple pressure of earning a living for his family made it impossible for van Goyen to pay off all his debts to van Ravensteyn, and when the burgomaster died in 1641 he still had not received most of his money. Even then the artist received no respite; van Ravensteyn's heirs continued to demand payment. The pressure on van Goyen proved so unremitting that his precarious finances collapsed into disorder and he was forced to arrange public auctions of his work on at least two occasions when he needed money in a hurry.

Jan van Goyen lived on until 1656, two decades after the collapse of the bulb craze which had ruined him, and he was still insolvent when he died. He left a substantial body of brilliant landscapes, many of which would probably never have been painted had he made his fortune in the tulip trade, and a debt which still totalled 897 guilders.

He was the last known victim of the tulip mania.

CHAPTER 14

At the Court of
the Tulip King

The final liquidation of the Dutch mania at the beginning of 1639 left many Hollanders with a distinct aversion to tulips. The episode did not entirely put off the wealthiest collectors of the rarest blooms, who continued to pay high prices for individual bulbs for another hundred years. But interest in tulips otherwise fell away in the United Provinces, now there was no possibility of making a quick fortune from the flower.

Yet the world had not seen the last of tulip mania. Like bubonic plague, it was a strange and complex disease which could rage for a while and then seem to disappear when – like plague – it was really only lying dormant. And, like plague, it could reappear miles away and decades later, as virulent as ever.

Thus it was in the Ottoman Empire. During the first half of the seventeenth century, the tulip lost some of its lustre for the Turks. The decline set in around 1595 with the accession of a womanizing sultan, Mehmed III, who was less interested in flowers than he was in seducing two or preferably three of the ladies of the harem each night. The rulers who followed in Mehmed's wake – from the fantastically misogynist Mustafa I, who ended his reign locked, as a sort of punishment, in a dungeon with only two naked female slaves for company, to the unfortunate Osman II, who suffered an agonizing death 'by constriction of the testicles' at the hands of his own soldiers – were on the whole either short-lived inadequates or butchers. They displayed at best only sporadic interest in the gardens of the Abode of Bliss.

It was not until the sultanate of Mehmed IV, who reigned from 1647 to 1687, that some degree of stability returned to the Ottoman Empire. Although his own father, Ibrahim the Mad (a libertine who once had all 280 women in his harem drowned simply so he could have the pleasure of selecting their replacements), was noted for his love of tulips, Mehmed was the first sultan in half a century to devote himself to horticulture in a significant way. It was he who first planted an imperial garden solely devoted to tulips in the Fourth Court of the palace, where it was to flourish for a century, and he who decreed that each new species of the flower should be registered and classified. To supervise this process, the sultan established a formal council of florists which sat in judgement on new cultivars, noted their special characteristics and assigned to the most flawless the poetic names beloved of the Turks – the Pomegranate Lances and the Delicate Coquettes. This council long outlived its master and continued to hand out verdicts on new tulips for another hundred years.

Unfortunately for Mehmed, his empire proved more

difficult to manage than his flowers. The last years of his reign were marked by a series of military disasters in the Balkans which severely weakened his authority. Worse still, bread prices in Istanbul quadrupled and led to unrest in the capital itself. At the end of 1687 the sultan's own ministers arranged for him to be deposed and replaced by a pliant half-brother.

There was a good reason why the Turks were cursed with the long line of mad or bad sultans who threatened to ruin the Ottoman Empire more or less throughout the seventeenth century. Things had changed in Istanbul since the days of Suleyman the Magnificent. Much of the vigour of the Turkish royal line had dissipated when it eventually proved necessary to abandon the old ways of securing the imperial succession. Ever since the time of Bayezid, the victor of Kosovo, the sultanate had gone to whichever royal prince could seize it first; and – following Bayezid's bloody example – the new sultan would inaugurate his reign by having every one of his brothers executed so they could not plot to usurp him. Under Mehmed the Conqueror, this lethal tradition had actually been codified as law, so that on the accession of Mehmed III in 1597 no fewer than nineteen of the new sultan's siblings, some of them still infants at the breast, had been dragged from the harem and strangled with silk handkerchiefs – having first been circumcised to ensure they would receive a welcome in paradise. Brutal as the system was, it produced a series of bold, decisive sultans famous for their ruthlessness. In 1607, however, the reigning sultan, Ahmed I, could no longer stomach the prospect of one of his beloved children murdering all the others. He arranged for the old policy of legal fratricide to be replaced with one of locking up unwanted brothers in a small area of the harem known as the *kafes*, the cage.

The cage was a suite of rooms to the west of the fourth courtyard of the palace which offered tantalizing views over fig orchards, the Ottoman paradise gardens and the Bosphorus. There, with eunuchs for company and sterile concubines for sexual consolation, unwanted princes lived lives which unpleasantly combined the immutable boredom of their daily round with the nagging terror that execution might, after all, still be their lot. When one Ottoman ruler died, his eldest son would be taken from the cage where he had spent his entire life and acclaimed as the new sultan, while the other men of the imperial line would return to the few pursuits they were permitted – embroidery and the manufacture of ivory rings among them – and their lives of quiet despair.

At the beginning of the eighteenth century, the succession devolved at last upon a son of Mehmed IV named Ahmed III, who had spent the first twenty-nine years of his life locked within the cage. Ahmed proved to be not only the most sophisticated and cultured sultan to have reigned since Suleyman the Magnificent, but also, without question, the greatest tulip maniac known to history. Having been inspired with a love for the imperial flower by his father, and having spent his days gazing longingly from the cage's marble balcony across the most gorgeous private gardens in the Ottoman Empire – gardens he was never allowed to wander through or touch – Ahmed came to the Ottoman throne suddenly equipped with almost unlimited means to indulge his whims.

The most avid bulb dealer of the Haarlem colleges could

hardly have competed with Ahmed's enthusiasm for tulips. The new sultan was besotted by the flower, so much so that the tulip became the most prominent feature of his long reign and the Turkish historian Ahmed Refik was moved to bestow the title *lale devri*, the Tulip Era, on the period. From the time of his accession in 1703, tulip mania burst forth again, in Istanbul this time. It was to rage on in the Turkish capital for almost three decades.

In truth, this time of tulips masked the uncomfortable reality that the great empire of the Ottomans was in decline. Turkish power was on the wane everywhere from the African littoral to the eternally war-torn Balkans, where the Peace of Karlowitz, signed in 1699, had ceded Hungary and Transylvania to the Austrians, ended the era of Ottoman expansionism in Europe, and pushed the imperial frontier back to within a few hundred miles of Istanbul. The flower festivals that characterized the Tulip Era, and the pomp that went with them, were distractions which the sultan's ministers ordered to divert their people from the realities of the political situation and their master from the tribulations of ruling an unwieldy empire.

In fairness, Ahmed was more than simply a tulip maniac. He fought the Russians with success, and was a builder and a bibliophile during whose reign Ottoman embassies – the first of their kind – were dispatched to the capitals of Europe to gather information and ideas from the West. He left, in the Ahmed III fountain which stands just outside the Topkapi Palace, one of the most gorgeous monuments to adorn the imperial capital. Nevertheless, he did preside over an era of hedonism unique even at the Turkish court. For three decades, the once warlike Ottomans gave themselves over to pleasure and disported themselves at numerous festivals – day or sometimes week-long cele-brations of music, feasting and elaborate ceremonial organ-ized by their monarch and his ministers. 'Let us laugh,' Ahmed's closest companion, the court poet Nedim, wrote in setting out the informal philosophy of the reign. 'Let us play, let us enjoy the delights of the world to the full.'

Free though he now was to enjoy the trappings of power, Sultan Ahmed found there were disadvantages to being the king of kings. He once complained that he had to dismiss no fewer than thirty-five of the privy chamber pages who routinely crowded into his bedchamber before he could feel comfortable changing the imperial trousers in front of the remaining three or four. But being sultan unquestionably had its advantages too. For the marriage of a favourite daughter, Ahmed had the palace confectioners spin edible sugar bowers, each 18 feet in length, in which the wedding guests could nibble at the foliage. On other occasions, guests wandered through gardens filled with jugglers, wrestlers, dwarfs and – an Ottoman speciality – silver *nahils*, artificial trees up to 60 feet tall, made of wax and wire and covered in mirrors, flowers and jewels.

Perhaps the most elaborate of Ottoman festivals were those that marked the ritual circumcision of the sultan's heirs. These were generally organized a year or more in advance, dragged on for weeks, and culminated in the presentation to the princes' mothers of golden plates bearing their sons' severed foreskins. In 1720, Ahmed III held such a festival to mark the circumcision of four sons and the marriage of two more of his daughters. It lasted for fifteen days and nights, and involved the construction of forty-four *nahils* for each young prince, the simultaneous circumcision of five thousand other Turkish boys, and the driving of carriages across tightropes slung between some of the ships which crowded into the Bosphorus to join the celebrations. But such affairs were necessarily rare. In the absence of more daughters to marry and more sons to circumcise, Ahmed and his ministers devoted much of their attention to annual tulip festivals held in the gardens of the Topkapi's innermost courtyard.

The tulip festivals took place in April, when the flowers were in bloom, and occupied two successive evenings during the full moon. They were purposely spectacular. On the first evening the sultan sat in state in a kiosk built within the garden and received the homage of his ministers to the accompaniment of songbirds chirping in an aviary suspended in the trees, while other guests – all strictly forbidden to wear clothes which did not harmonize with the flowers – wandered through tulip beds illuminated by candles fixed to the backs of slow-moving tortoises. On the second evening the male guests were banished while the sultan entertained the ladies of the harem and held treasure hunts among the flowers. Sometimes the prizes were candies; sometimes, precious stones. At the end of each evening's entertainment, the chief white eunuch – a Christian slave who acted as palace chamberlain while his Abyssinian colleague, the chief black eunuch, took charge of the harem – distributed gifts of robes and jewels and money to those who basked in the sultan's favour.

Ahmed's passion for tulips – not the varieties that the Dutch had coveted, but slender, needle-pointed Istanbul tulips – was such that the flower soon found new favour among all classes in the capital. Barbers and shoemakers cultivated bulbs. So did the *sheikh-ul-islam*, the most senior cleric in the Ottoman Empire. Demand for the finest tulips was great – a single bulb of the cultivar Mahbub ('Beloved') could change hands for as much as 1000 gold coins – but, perhaps learning a lesson from the Dutch, Ahmed averted a trading mania by limiting the number of florists who were permitted to operate in the capital and by fixing the prices of the most coveted blooms by imperial decree. Even firmer measures were taken to damp down speculation in the Ottoman provinces. Eventually it became a crime punishable by exile to sell tulip bulbs outside the walls of Istanbul.

Centuries of effort had produced a startling diversity of tulips by Ahmed's day. One of the official price lists fixing the value of the best-known cultivars named more than 820 varieties, and fresh hybrids of tulip continued to be developed throughout the reign. Such was the interest in the flower that the first appearance of a new cultivar was often memorialized in poems known as chronograms, which recorded the auspicious date in the letters of the final verse.

In important respects, though, it was all too little, too late. The neglect which the tulip had suffered during much of the seventeenth century meant that by the time Ahmed came to the throne the Ottomans had long lost their primacy in the cultivation of the flower, and the Turks, who had given the tulip to Europe, now imported thousands of bulbs each year from the Netherlands and France. Nevertheless, the Ottomans retained rather fixed ideas as to what precisely constituted an ideal flower. *Mizanu 'l-Ezhar, The Manual of Flowers*, a manuscript written by Ahmed's chief gardener Seyh Mehmed Lalezari, lists twenty criteria for judging a tulip's beauty. The stem should be long and strong, Seyh Mehmed wrote, and the six petals smooth, firm and of equal length. The leaves should not hide the blossom, however, and blossom should stand erect; nor should the flower be soiled with its own pollen. Variegated flowers should display their colours on a pure white background.

This stark description scarcely does justice to the uniquely poetic flavour of the Ottoman desiderata, however. Another of Lalezari's manuscripts, which survives in an archive in Berlin and bears the title *Acceptable and Beautiful*, describes the ideal tulip as

> curved as the form of the new Moon, her colour is well apportioned, clean, well proportioned; almond in shape, needle-like, ornamented with pleasant rays, her inner leaves as a well, as they should be, her outer leaves a little open, as they should be; the white ornamented leaves are absolutely perfect. She is the chosen of the chosen.

One may be quite certain that the rare species which met these exacting criteria would have found their way to Ahmed's gardens.

The sultan's servants soon found it expedient to share his passion for the flower, and many became considerable enthusiasts in their own right. Mustafa Pasha, the admiral of the Ottoman fleet, created forty-four new varieties. Ambitious minor officials discovered they could bribe their way to high office with presents of fine tulips. Nor was it wise to deny the king of kings a flower he particularly coveted. When one rare bulb – the gift of a canny European ambassador – went missing, the Grand Vizier of the day packed heralds off into Istanbul's narrow streets to offer huge rewards for its safe return.

Early in his reign, Ahmed III had followed the example of his immediate predecessors by working his way through a succession of short-lived viziers. Fazil Pasha was one: honest, hard-working, last scion of a distinguished line of imperial servants, but also an eccentric who believed there was a fly perched on the end of his nose, which returned each time he brushed it away. In 1718, however, the sultan appointed a man named Ibrahim Pasha Kulliyesi as Grand Vizier of the Ottomans. Ibrahim was a shrewd manipulator of imperial intrigue who made it his business to forge the closest possible relationship with the sultan. His greatest coup was his marriage to Ahmed's eldest daughter, which earned him the nickname Damat ('son-in-law'). In a land where the office of vizier had long been synonymous with short tenure and an often violent death, Damat Ibrahim clung to power for a dozen years.

The son-in-law's policy was one of cautious progress – just what the declining but still intensely conservative empire required. It was Ibrahim who induced Ahmed to send Turkish embassies to learn about advances in the West, Ibrahim who set up the first Ottoman fire brigade, and Ibrahim who licensed an official printing press to produce books on science and geography. He levied new taxes, restocked the imperial treasury, and kept most of the empire at peace. Most importantly, however, the Grand Vizier retained the favour he needed to push through his programme of reform by indulging Ahmed's love of fine flowers.

The French ambassador, Jean Sauvent de Villeneuve,

described one royal entertainment, held in Ibrahim's own tulip garden:

> Beside every fourth flower is stood a candle, level with the bloom, and along the alleyways are hung cages filled with all kinds of birds. The trellises are decorated with an enormous quantity of flowers, placed in bottles and lit by an infinite number of glass lamps of different colours. The lamps are also hung on the green branches of shrubs which are specially transplanted for the fête from the neighbouring woods and placed behind the trellises. The effect of all these varied colours, and of the lights which are reflected by countless mirrors, is magnificent.

The illuminations, Villeneuve added, continued nightly at Damat Ibrahim's personal expense, 'so long as the tulips remain in flower'.

Using his ambassadors' rapturous reports of the French royal palace at Fontainebleau and Louis XV's château at Marly as his guide, the Grand Vizier built a villa for himself in a quasi-European style. It stood on the Bosphorus just above Istanbul, and when Damat Ibrahim entertained Ahmed there in the spring of 1721, the impressed sultan immediately ordered the construction of a new royal palace in similar style nearby. The palace chosen was a spot where two streams known as the Sweet Waters of Europe ran through meadows down to the sea. Here Ahmed's architects constructed a sumptuous pleasure palace called the Sa'adabad ('the Place of Happiness'). It took them just three months, in the summer of 1722. Perhaps for the first time in the Ottoman Empire, the gardens were planted in the more formal European style, with avenues of trees leading to square and regimented beds of tulips. The Sweet Waters themselves were turned into marble-banked canals which fed fountains and cascades surrounding a central ornamental lake.

By keeping the people of Istanbul supplied with cheap bread, and the sultan sated with festivals, Damat Ibrahim remained in office throughout the 1720s. But eventually even he ran out of luck. Events far beyond the gardens of the Sa'adabad were moving beyond his control. Ruinous taxation, necessary to fund not just the ostentations of the court but also a war against the Persians which erupted in the early 1730s, combined with famine to set the imperial provinces in turmoil. Worse, Ottoman armies were soon falling back in disarray on the eastern frontier and the hated Persians recovered substantial swathes of land the Turks had seized from them earlier in the century. When news of these imperial defeats reached Istanbul, the mutterings of discontent that had been circulating in the bazaars turned into outright demands for change.

Not even the Grand Vizier could keep such bad news from reaching the sultan; not even Ahmed III could afford to ignore it. By the time the Istanbul mob – led by an Albanian second-hand clothes dealer called Patrona Halil – marched on the Topkapi in the autumn of 1730 clamouring for scapegoats, Ahmed knew that his reign was in grave danger of ending prematurely and that if he failed to placate the crowd his own life might be forfeit. In this sudden crisis, expediency was all to the tulip king and he ordered his corps of gardener-executioners, the *bostancis*, to surrender up the heads of Damat Ibrahim and Mustafa Pasha, the ministers most closely associated with the unpopular policies of Westernization and reform.

It was the beginning of the end – for Ahmed and for the Tulip Era too. The Grand Vizier was discovered at his official residence, strangled and decapitated. Then the *bostancis* set off for Mustafa's waterfront villa near the Sa'adabad. They came upon the Grand Admiral transplanting tulips in his garden, blissfully unaware of the sudden political crisis in the capital. Perhaps the gardeners who had come to kill the pasha paused for a moment, while their victim prepared himself for death, to cast a professional eye over the forty new varieties of tulip he had created. But if they did, they certainly could not have known that, as the silken bowstring tightened around the Grand Admiral's neck and Mustafa began the journey from his paradise-garden to the gardens of paradise, the time of tulips was all but over.

Ahmed had done too little, and acted far too late, to save his throne. The mob would not disperse, and the sultan's position soon became critical. Perhaps a more resolute monarch, one whose skills ran more to military matters than to organizing tulip festivals, might still have rallied some loyal troops and saved himself. But Ahmed was no general and he survived the sacrifice of his closest advisers by only two days. As rioting engulfed Istanbul and control of the capital slipped from his grasp, the sultan was persuaded that his only chance of saving his own neck was abdication.

A nephew, Mahmud, was plucked from the cage and placed on the throne in Ahmed's place. His accession was a turning point for both the empire and the tulip, for though he soon dealt ruthlessly enough with the rioters who had deposed his uncle and run wild through Istanbul, burning the wooden tulip kiosks which had symbolized Ahmed's reign, the new sultan's real interests lay elsewhere. He was a keen voyeur who asked nothing better than to hide behind a grill in the harem and spy on the women of the palace. On one occasion the sultan even had the stitches of the flimsy clothes the ladies wore while bathing secretly removed and the garments reassembled with glue, knowing that it would melt in the heat of the steam room and expose each woman, naked, to his gaze.

Such a monarch could never accord a mere flower the exaggerated respect it had enjoyed in the time of Ahmed III. Though flower festivals continued to be held each spring, they were far more modest than they had been during the *lale devri*, and the tulip's second decline in Turkey dates from the reign of Mahmud I. In the end it was so complete that the whole gorgeous panoply of Istanbul tulips – all 1300 varieties and more – slowly vanished from the gardens of the empire and the memories of men. Today, not a single specimen survives.

And what of the sultan who presided over the tulip's late flowering? He was permitted to live, but only after a fashion. Ahmed, the tulip king, was returned to his cage to gaze once more over the Ottomans' fig groves – and while away his nights in dreams of dagger-petalled flowers bathing in the full moon's light, and throwing needle-pointed shadows about the secret gardens of the Abode of Bliss.

CHAPTER 15

CHAPTER 15

Late Flowering

That was the end of the tulip mania. When the door of the Ottoman cage slammed shut for the last time on Ahmed III, the flower began to fade from the history books. Its greatest years were past; never again would it so captivate a king, or enslave half a nation with the promise of easy money. In time, people would come to wonder how such a mania could ever have occurred at all.

But if the tulip ceased to be a public craze, it remained a private passion. It would be quite wrong to suppose that the collapse of the bulb trade put paid to all interest in tulips, or even that prices plunged and then remained uniformly low in reaction to the excesses of the Dutch and the Turks. On the contrary, quite large sums were still demanded for the bulbs of a very few rare and highly regarded varieties.

It took only a year or two for the Dutch bulb trade to regain some sort of equilibrium. The speculators had gone, but there was still a market for the flower; the purchasers were the same patrician collectors who had stayed aloof

from the tavern trade, and they continued to value the tulip
for purely aesthetic reasons. Even in the summer of 1637,
less than six months after prices crashed, a Haarlem connoisseur named Aert Huybertsz. paid 850 guilders for a
single bulb of the fine *Rosen* variety Manassier. Jacques
Bertens, the dealer he purchased the tulip from, had earlier
paid 710 guilders for the flower and thus profited to the
tune of 140 guilders, or about six months' wages for a local
artisan.

The fashion among tulip connoisseurs in the post-mania
years was to cultivate single specimens of as many different
tulips as possible. This meant there was still at least a limited
demand for many of the most attractive varieties of flower.
In an odd way, the infamy that the mania had attracted
helped too; the whole of Europe had heard of tulips now,
and many people wanted to see for themselves the flower
that had generated such passions. Dutch growers were thus
able to offset their domestic catastrophe by developing an
export business. A fair number enjoyed considerable success;
indeed the dominance that Holland still enjoys in the
international flower trade dates to the first half of the
seventeenth century.

This steady business was of inestimable value to the
gardeners of Haarlem, who certainly must have lost a good
proportion of their customers to the mania, and from
scattered hints it appears that the bulb growers did what
they could to keep the supply of the most favoured species
low. They thus contrived to maintain prices at a decent level
for years, cannily resisting the temptation to breed more
tulips and risk flooding the limited market that remained.

Comparatively little data concerning the prices paid for

tulips has survived for the years after 1637. Peter Mundy, who travelled through the United Provinces in 1640, noted that 'incredible prices' were still being paid for what he called 'tulip rootes', without giving examples. But the sort of sums Mundy, a reasonably well-off merchant, would have considered incredible were still far short of those commanded in 1636 and 1637. An Admirael van der Eijck, which sold for an average of about 1345 guilders per bulb at Alkmaar, went under the hammer for only 220 guilders when another grower's estate was auctioned off in 1643, and a Rotgans once worth 805 guilders for only 138. Without knowing the precise weights of the bulbs concerned it is impossible to say for certain that the sums are truly comparable, but in both these cases prices had fallen to only a sixth of what they had been at the height of the mania, an average annual depreciation of 35 per cent.

If the rarities fared badly, then – as might be expected – the cheaper bulbs did far worse. They had appreciated late, and only when the stock of more desirable bulbs seemed exhausted; they were too common and too drab to interest the connoisseurs. Witte Croonen – plain White Crowns – which had sold for 64 guilders per half-pound in January 1637 and then rose to the giddy heights of 1668 guilders the half at Alkmaar, could be had for only 37 guilders, 10 stuivers five years later. To reach that low, they had depreciated at a spectacular average of 76 per cent per year.

Such prices were not enough to sustain everyone who had dabbled in bulb growing. In the years that followed the mania, the fledgling flower industry contracted and most of the new and inexperienced growers who had been attracted by the prospect of rich profits gave up the business or were driven out. Tulip breeding retreated quite literally to its roots in the rich sandy soils around Haarlem; indeed, the town now established a total dominance over the bulb trade such as it had never enjoyed when the times were good and everyone was growing tulips. By the end of the seventeenth century even the Turks had yielded their old primacy in the tulip trade to the nurseries of the city, and during the reign of Ahmed III, the farms of Haarlem shipped tens of thousands of bulbs to the Ottoman court in Istanbul. The town became so closely associated with the finest flowers that the handful of florists who did base themselves away from the city routinely listed their address as 'Near Haarlem' when they sent out catalogues and price lists. They knew their produce would be dismissed as second-rate if they did not.

The trade was much more rational now. The bulbs that did command high prices were taken to the auctions which continued to be held in Haarlem for the remainder of the seventeenth century. The tulips sold at these affairs would have been those of new varieties, recently developed and still rare enough to command a premium. After a few years, most of these newcomers would lose their sheen, and the connoisseurs would move on to other novelties. In time, the once-fashionable bulbs would become relatively commonplace and the growers would begin to sell them to callers or via mail order through garden catalogues aimed at more modest purses. It would appear from surviving lists of the extensive bulb purchases made by one German tulipophile – Charles, Margrave of Baden-Durlach – that by about 1712 the bulbs available from these catalogues cost only a guilder apiece on average, although a few varieties might command 10, 20 or even 40 guilders a bulb. The number of species and the number of bulbs available by the next century was much greater, too. An inventory of the Margrave's collection showed that in 1736 he owned not only 4796 different varieties of tulip, but as many as 80,000 bulbs of a single species.

Preferences had not changed much, and the Margrave's flowers would have been recognizable descendants of the tulips grown during the years of mania. The mosaic virus remained undiscovered, and brightly coloured, flamed tulips continued to be highly popular. Indeed the desiderata applied to the most coveted varieties would have been perfectly familiar to a flower dealer of the 1630s: in 1700 Henry van Oosten's* *The Dutch Gardener* noted that the ideal tulip should have

> petals that are rounded at the top, and these should not be
> curled ... as for the flames, these must start low, beginning
> at the base of the Flower and climbing right up the Petal,
> and ending in the form of a shell at the edge of the flower
> ... As regards the base, it must be of the finest Sky Blue,
> and the Stamens should seem to be Black, although they are
> really of a very dark Blue.

The Dutch Florist, a book by Nicholas van Kampen translated into English in 1763, added that the 'properties required of

---

* At least, that was how he was known in Britain. His proper name, Henrik van Oosting, was evidently too much of a mouthful for his British publisher.

a fine tulip' were a tall stem, a well-proportioned cup and lively colours, preferably on a background of white.

Even so, no plant, not even the tulip, could hope to remain in fashion for ever. Tastes changed; other flowers offered something different. Although the French, in the eighteenth century, and the English, in the nineteenth, retained a passion for the flower, the tulip was often relegated to the second rank as other species came briefly into vogue and occasionally generated miniature manias in their own right.*

* There are close parallels between the flower mania and crazes for other commodities. Similar booms – by which economists mean exceptionally rapid rises in prices – and bubbles (booms in which a commodity's price quite outstrips what it is actually worth to anyone other than a speculator) have occurred all over the world throughout the last 400 years. The objects of speculation have varied from the obvious – stocks and shares, land, oil – to the unusual. In the United Provinces themselves there was a boom in the passenger canal system begun in 1630, a genuinely useful development in the transport system which made many men rich, and, during the 1670s, a bubble involving the erection of elaborate public clocks, which Dutch merchants built as status symbols.

Of all bubbles, however, the one which perhaps resembles the tulip mania most closely was the Florida Land Boom of 1925. Like the tulip, Florida was exotic but, before 1925, the state was difficult to get to and both unhealthy and swampy. Gradually, however, the construction of new roads and railroads and the draining of the swampland, together with the guarantee of fine winter weather, made it more attractive, and some rich Americans invested in holiday homes in the Miami area. Poorer people were attracted by their example, and local real estate agents were quick to exploit the rising demand for property.

Stories began to circulate concerning the fantastic profits which could be made by buying and selling land in Florida. The famous lawyer William Jennings Bryan bought a winter home in Miami in 1912 and sold it in 1920 for a profit of $250,000. Later on, lots purchased for $1,200 could be resold a few months later for $5,000. A lot acquired for $2,500 was resold for $7,800, then $10,000, $17,500 and finally $35,000 – the last purchaser being the man who had sold it for $2,500, and had lived to regret doing so. At Snapper Creek Canal, land worth $15 an acre in 1913 sold for $2,000 an acre in 1925, and in central Miami land once

Perhaps the most striking of these affairs was the hyacinth trade which grew up in the United Provinces in the first third of the eighteenth century. Like the tulip, the hyacinth was introduced to western Europe from the Ottoman Empire in the sixteenth century. Clusius knew it and distributed its bulbs, and it was cultivated in the Netherlands in a minor way for several decades without arousing any great passion among flower lovers. Then chance intervened. Over the years, growers trying to create new varieties had accidentally produced a few double hyacinths – flowers with twice the usual number of petals. Because these plants did not produce seeds, they were routinely destroyed, and hyacinths occupied a station in the florists' pantheon below

---

worth $30 per acre became worth $75,000. Eventually, land in Miami became more valuable than property on Fifth Avenue, New York. Much of it was bought on payment of a small deposit by speculators who planned to resell it before their next payments became due.

Money poured into the state. In a twelve-month period beginning in the autumn of 1924, bank clearings in Miami rose from $212,000 to over $1m, and land transfers tripled. An edition of the *Miami Daily News* published in the summer of 1925 ran to 504 pages, almost all of it real estate advertising – a world record at the time. It was said there were 2,000 estate agents in Miami alone, employing 25,000 salespeople.

The crash came in autumn, as crashes often do. Speculators had badly overestimated the real demand for land. The number of winter visitors to the state was only a tenth of what had been predicted. People began to default on their loans, and a man who had sold land for $12 an acre and seen successive purchasers pay $17, $30 and $60 an acre was dismayed to discover that all had failed to pay more than their initial deposit, leaving the land to revert to him. From the summer of 1926, the crisis had caused several Florida banks to fail as clearings in the state dropped from $1bn in 1925 to $633m a year later and eventually to a mere $143m in 1928. In the latter year, *The Nation* wrote, 'Miami will be the cheapest place in the United States to live ... One of the most pretentious buildings on the beach, whose monthly rate was $250, now rents for $35 ...'

that of the tulip and the carnation. In 1684, however, a Haarlem bulb farmer called Pieter Voorhelm fell ill and was unable to tend his garden for some time. When he recovered and went to dispose of some double hyacinths he had been meaning to get rid of, he discovered that an especially fine double had flowered and that some of his customers wanted to buy it. Not only that; they were willing to pay more for the new flower than they were for the single hyacinths he produced.

Voorhelm continued to grow the new variety and, as demand slowly increased, he bred more doubles. Other growers followed suit, until by about 1720 the hyacinth was definitely in fashion and had quite eclipsed the tulip in popularity.

The craze that ensued bore strong similarities to the tulip mania, and even ran its course more or less exactly a century after tulips were in vogue. It began slowly and did not reach a peak until 1736, half a century after Voorhelm first grew a double hyacinth. Relatively early on, the prices for single bulbs of the most prized species reached 30 or 40 guilders, and before the fashion had run its course, the Semper Augustus of the hyacinth years – a double named Koning van Groot Brittannië in honour of William of Orange – was fetching 1000 guilders a bulb.

Hyacinths were popular for exactly the same reason that tulips had captured the imagination. It took a similarly long time – five years – to produce a flowering bulb, which meant that popular new hyacinths remained rarities for some time. The new varieties were highly variegated, exhibiting endless combinations of colour mainly in shades of blue and violet, and so beautiful that one dealer, Egbert van der

Vaert, used to boast that if Zeus had only known of his latest acquisition he would have taken on the form of that hyacinth, rather than a swan, when he descended from Olympus to seduce Leda. And, unlike the tulip, they were heavily and exotically scented.

During the 1720s, then, bulb prices began to rise. In one sense this was odd, because the cultivation of bulbs was a considerably more professional business in the eighteenth century than it had been a hundred years earlier, and new varieties of double hyacinth soon began to flood on to the market – a total of two thousand were eventually produced. That might have sated demand and prevented a real mania from developing. But the bulb growers of Haarlem had accumulated a better understanding of their business, too, and knew they could push their profits up by keeping the supply of the most favoured bulbs low.

By 1730 hyacinth prices had reached substantial levels, much to the delight of the growers. The Voorhelm bulb gardens, run now by Pieter's grandson Joris, remained at the forefront of the trade, but other Haarlem growers also made fortunes from hyacinths. Prices peaked between 1733 and 1736 before falling away steeply in 1737. The reason for the plunge was the same that it had been in 1637: prices had reached such high levels that the most desirable bulbs became all but unobtainable and less fancied varieties appreciated to the point where they cost far more than they were worth to any real flower lover. Bulb catalogues published two years after the mania's peak show that valuable doubles such as the white Staaten Generaal, which had sold for 210 guilders, now fetched only 20; Miroir went from 141 guilders per bulb to 10, Red Granaats fell from 66 guilders to 16, and Gekroont Salomon's Jewel from 80 all the way down to 3.

From these figures it can be seen that the prices obtained during the hyacinth craze were of an order of magnitude less than those of the tulip craze. Staaten Generaal sold for around 200 guilders, where an Admirael van der Eijck might have fetched nearly 2000, and the highest prices recorded for double hyacinths, at about 1600 guilders per bulb, were at best only a third as much as the most coveted tulips had fetched a century before. In addition, individual speculators appear to have been a little more cautious than their forebears. The one significant innovation of the hyacinth craze was the widespread practice of buying shares in particularly valuable bulbs. It must have been a frustrating business, in that the shareholders would have to wait a year or more for their flower to produce offsets before they could expect to receive a single bulb of their own, but it was at least a cheap way of buying into hyacinths; one lengthy Dutch poem, *Flora's Bloemwarande*, which described the new trade, mentions a florist named Jan Bolt, who sold a half share in one of his bulbs to a hesitant customer, with only 10 per cent down.

There were several reasons why the hyacinth trade never matched the tulip mania in magnitude. To begin with, hyacinths are much more difficult to grow than hardy mountain flowers such as tulips, which limited the number of garden lovers interested in buying them. This in turn meant that demand remained at a lower level than it did during the years of tulip craze; hyacinths attracted much less attention than tulips had done, which kept the number of speculators attracted by the trade to a minimum. Most significantly of all, there is little evidence of any sort of futures trade in hyacinths; there are one or two mentions of bulbs being purchased and then sold on to third parties, but nothing more.

Nevertheless, at least a few private enthusiasts in Haarlem and The Hague seem to have been sufficiently caught up in the hyacinth craze to attempt to grow the flowers themselves for profit, and at its peak there was considerable disapproval for the new fashion. Memories of the tulip mania evidently remained vivid, for one enterprising publisher reprinted the three *Samenspraecken* of Gaergoedt and Waermondt, prefacing the dialogues with the comment that the present-day speculators were just as greedy as their ancestors, and just as taken in by tawdry deceits of that wily old whore, Flora. Others produced new tracts warning against the excesses of the hyacinth trade. With the awful lessons of the tulip mania so fresh in every mind, it might be said that the most remarkable thing about the new craze was that it occurred at all.

The story of the tulip can be brought up to the present day in a very few words. The trade has continued to be dominated and driven forward by Dutch growers. Indeed, for much of the eighteenth century a single group of a dozen Haarlem florists effectively controlled the entire business. Even when their oligopoly was broken during the Napoleonic Wars, the reputation of Dutch farmers remained unparalleled, and as more and more people took up the garden as a hobby and worldwide demand for flowers of all sorts soared, the area around Haarlem given over to the cultivation of bulbs increased too. First farms appeared in

Bloemendaal and Overveen, just to the west of the city; then cultivation expanded south towards Hillegom and Lisse on land made available by the draining of the Haarlemmermeer in the middle of the nineteenth century. It was around this time that individual bulb farms expanded in size too, creating the first of the huge tulip fields that have become one of the most popular picture-postcard images of Holland. Next – with almost all the fertile land around Haarlem given over to flowers – a portion of the bulb trade moved away from the city altogether. Today, more tulips are produced by the farms of north Holland than come from Haarlem.

There have been other fundamental changes, too. Bulb growers have now mastered the techniques necessary to produce tulips all the year round. By keeping bulbs at low temperatures in a state of suspended animation, it is possible to have them flower as desired. The long wait for the next tulip time, which frustrated flower lovers for centuries, no longer exists, and with it has vanished the single most essential precondition of the tulip mania.

Most fundamentally of all, the tulip itself has changed. In the 250 years that have passed since the mania subsided, Dutch farmers have introduced several radically different species to gardens, from parrot tulips, with their twisted leaves and big, beak-tipped petals, and double tulips, with their extra complement of petals, to Darwins – hybrid giants first bred in the nineteenth century. The broken tulips that once achieved such fame, on the other hand, are now all but extinct. Weakened as they were by the mosaic virus, the original species – including even famed varieties such as Viceroy and Semper Augustus – were in any case doomed to flourish for only a short time, but even their successors are long gone now. A few English tulip clubs still possess broken varieties, but for years the only flared and flamed tulips available to other gardeners have been imitations produced by careful cross-breeding.

The bulb industry views the destruction of the mosaic virus as one of its proudest achievements, with good reason. It is the florists' equivalent of the elimination of smallpox. Yet it can hardly be denied that something has been lost in the winning of this war. The infinite variety that each broken tulip could display is gone, and with it much of the flower's capacity to fascinate and astound.

Today, the bulb trade offers not variety but varieties: a huge and ever-growing array of different tulips. The flower lover of Clusius's day had only a handful of species to enjoy, but now close to six thousand different tulips have been bred, described and catalogued.

This dazzling array of choice is certainly impressive in itself; yet it unarguably lessens the importance of individual flowers. The modern fashion for expanses of uniform and unicoloured tulips would certainly strike the seventeenth-century connoisseur, with his exemplars planted in their own small beds, as rather vulgar; and surely no modern gardener studies his flowers with the intensity of an old-time tulipophile, or knows each one so well.

As for tulip mania – well, that is one virus that has never disappeared. It always was a purely human disease, one which fed on the complementary human emotions of appreciation of beauty and greed for money, and it still breaks out occasionally. There was, for example, a craze for dahlias in France around 1838. Like the tulip two centuries before, this flower was a relative newcomer to Europe, having been introduced from Mexico around 1790. It was soon taken up by horticulturalists, who bred numerous new varieties, and the beauty of the new cultivated flowers won widespread acclaim; they were even cited to disprove Rousseau's contention that in the hands of man everything degenerates. For a short while dahlias fetched high prices; a bed of the flowers is said to have changed hands for 70,000 francs,* and a single dahlia was exchanged for a fine diamond. Then fashions changed and the dahlia, like the tulip, faded from the history books. In 1912, it was the turn of Dutch gladioli to enjoy a very similar – but equally short-lived – boom.

The most recent manifestation of the old virus occurred as recently as 1985, when a mania broke out in China which followed the template of the tulip craze almost exactly. In this case, speculation centred on yet another bulbous flower, the *jun zi lan* plant, or *Lycoris radiata* – the red spider lily. This lily grows small, funnel-shaped flowers which coil together like a tangled skein of wool. Tremendously long, curved stamens project far beyond the leaves to give the plant a delicious air of delicacy. The spider lily originated in Africa, but came to China in the 1930s and was cultivated extensively in the Manchurian city of Changchun. It was at first a favourite of the old ruling classes of the city, and for a while it was a mark of distinction for a patrician family to grow several different varieties of *jun zi lan*. The Communist takeover put a stop to the small market for bulbs which had evolved by the end of the 1940s, but the spider lily remained very popular and was eventually designated the

* The equivalent of £12,000 at present-day prices.

official flower of Changchun. By 1980 it was estimated that half of all the families in the city grew it.

A *jun zi lan* mania broke out in earnest only a few years later, when the Chinese government allowed a few modest economic reforms. The situation in Changchun was then quite similar to that in Holland during the 1630s. Entrepreneurial activity was encouraged, but while there was plenty of desire to make money and an abundance of energy to tap, there were very few opportunities to invest any surplus cash. In these circumstances, the spider-lily growers of the city took advantage of increasing demand for their flowers from neighbouring regions, and as prices began their inevitable rise,* speculation in *jun zi lan* bulbs followed right behind.

In 1981 or 1982, spider-lily bulbs were selling for 100 yuan, about £15. This was already a substantial sum, given the low annual salaries prevalent in China. But by 1985, bulbs of the most coveted varieties are reported to have changed hands for the astronomical amount of 200,000 yuan, or about £30,000, an amount that puts even the sums paid at the height of the Dutch tulip craze to shame. Thus, while Semper Augustus at its peak might have commanded between 5000 and 10,000 guilders a bulb, which was four to eight times the income of a well-off merchant, the highest prices quoted during the *jun zi lan* mania were equivalent to no less than 300 times the annual earnings of the typical

---

* Even the most common and mundane objects can become rare and costly in certain circumstances. During the Second World War, when military supplies naturally took priority, US servicemen would go to great lengths to obtain bottles of Coca-Cola. On one occasion a single bottle of the drink, worth five cents, was auctioned on the Italian front for $4000.

Chinese university graduate; a quite staggering sum.

In such circumstances, it is unsurprising that the spider-lily craze was short-lived even by the standards of flower manias. It collapsed in the summer of 1985, apparently because confidence in the fledgling trade had been undermined by a series of critical newspaper articles which described the speculation in bulbs as madness.

The whole lily-bulb market was quickly flooded with panicked dealers desperate to sell, and bulb prices fell sharply. Just as the Chinese boom had exceeded even the heights attained during the tulip years, so the crash, when it came, was still more severe. By the time the market for spider lilies stabilized at last, prices had plunged by anything up to 99 per cent.

Changchun is in northern China, just north of the fortieth parallel and only 2000 miles from the valleys of Tien-shan. The mania virus had come home at last.

Notes

GENERAL

A surprisingly large amount is known about the history of the tulip, which enjoyed the good fortune of being highly regarded and of flourishing when garden writing was at its early apogee. As well as good early summaries such as Sir Daniel Hall's *The Book of the Tulip*, several scarce but excellent regional studies have appeared, notably Michiel Roding and Hans Theunissen's *The Tulip: A Symbol of Two Nations* and Sam Segal's pamphlet *Tulips Portrayed: The Tulip Trade in Holland in the Seventeenth Century*. The most comprehensive general account, however, is undoubtedly Anna Pavord's *The Tulip*.

Those interested in the history of the Netherlands in the seventeenth century are also richly catered for, most recently with the publication of Jonathan Israel's highly acclaimed overview *The Dutch Republic: Its Rise, Greatness and Fall, 1477–1806*. Social historians have Simon Schama's rather more controversial *The Embarrassment of Riches: An Interpretation of Dutch Culture in the Golden Age* and A. T. van Deursen's comprehensive *Plain Lives in a Golden Age: Popular Culture, Religion and Society in Seventeenth Century Holland*.

The history of the tulip mania itself, however, remains remarkably obscure, and even now it has never been the subject of an exhaustive scholarly enquiry making full use of the mass of raw material available in Dutch archives. Many of the short accounts of the subject are based on badly flawed popular studies, most notably Charles Mackay's entertaining but misleading *Memoirs of Extraordinary Popular Delusions and the Madness of Crowds*, which remains in print today despite having originally appeared in 1841. (Much more reliable, though still dependent on secondary sources, is the fairly extensive modern re-analysis of Joseph Bulgatz, published in *Ponzi Schemes, Invaders from Mars and More Extraordinary Popular Delusions and the Madness of Crowds*, which has, however, attracted very little attention.)

Apart from contemporary pamphlets, collected by E. H. Krelage in *De Pamfletten van den Tulpenwindhandel 1636–1637*, the most valuable Dutch sources are the Solicitors' Acts, which still exist for most of the cities caught up in the mania and record not only some of the (comparatively rare) legal agreements for the purchase of tulip bulbs but also the proceedings brought as a result of the collapse of prices in 1637. The extracts which have appeared, most notably those collated by A. van Damme, *Aanteekeningen Betreffende de Geschiedenis der Bloembollen: Haarlem 1899–1903* (a collection of turn-of-the-century journal articles finally gathered together and published at Leiden in 1976) and Nicolaas Posthumus, who published both pamphlets and some contemporary source material in 'De speculatie in tulpen in de jaren 1636 en 1637' parts 1–3, in the *Economisch-historisch jaarboek*, are in no way comprehensive; van Damme even states that the acts he published were chance discoveries rather than the products of systematic research.

By far the most exhaustive account of the period remains Krelage's monumental *Bloemenspeculatie in Nederland: De Tulpomanie van 1636–37 en de Hyacintenhandel 1720–36*, upon which a good portion of the present book is based. It is, however, now in some respects outdated. My general feeling, after reviewing the material available, is that even after sounding the necessary notes of caution about the reliability of the popular accounts, historians and particularly economists remain guilty of exaggerating the real importance and extent of the tulip mania.

INTRODUCTION

Success synonymous with virtue Paul Zumthor *Daily Life in Rembrandt's Holland* p.317.

'Gathered around the campfire ...' Peter Garber, 'Tulipmania', *Journal of Political Economy* 97 (I), June 1989 p.535.

PROLOGUE: A MANIA FOR TULIPS

The principal source of information on events in Alkmaar in February 1637 is A. van Damme, *Aanteekeningen Betreffende de Geschiedenis der Bloembollen*. On the appearance and behaviour of Dutch tulip traders, see both Zumthor and the more recent, and more analytical, A. T. van Deursen, *Plain Lives in a Golden Age*.

Breakfast Geoffrey Cotterell, *Amsterdam: The Life of a City* p.48.

Value of a tulip Peter Garber, 'Tulipmania' p.537n, states that in 1637 each guilder contained 0.856g of gold. One gramme of gold was thus worth 1.17 guilders. A Viceroy bulb sold at auction in Alkmaar on 5 February fetched 146 guilders per gramme, making it worth 125 times its weight in gold.

Richest man Jonathan Israel, *The Dutch Republic* p.348.

Tulip fortunes Garber p.550.

1: THE VALLEYS OF TIEN-SHAN

The early history of the tulip is very largely obscure. Its Asian origins are discussed by Turhan Baytop, 'The tulip in Istanbul during the Ottoman period', in Michiel Roding and Hans Theunissen (eds), *The Tulip: A Symbol of Two Nations*, and the enthusiasm for wild tulips in Persia rather briefly by Wilfrid Blunt, *Tulipomania*.

Asian origins of the tulip Baytop pp.50–6.

Early appreciation of tulips Certainly the Hittites, who dominated much of Asia Minor two thousand years before the birth of Christ, already appreciated the beauty of wild bulbous flowers. Ancient inscriptions record that the advent of spring was marked each year in the Hittite realm by a celebration called the *An.tah.sum-sar*, which may be translated as 'bulb festival' and which appears to have coincided with the first flowering of the crocus. (Today many Anatolians still celebrate a similar festival, called *Hidrellez*, each May, during which they go on picnics and eat a couscous of bulgar wheat and mashed crocus bulbs.) Possibly the flowering of the tulips held a similar significance for the peoples of the steppe, who experienced winters harsher than anything encountered in the crocus country of Asia Minor, and among whom the arrival of spring must have been at least as eagerly anticipated. See Baytop p.51.

The tulip in Persia Daniel Hall, *The Book of the Tulip* p.44; Blunt pp.22–3; Valerie Schloredt, *A Treasury of Tulips* p.62.

History of the Turks The Ottoman portion of the tulip's story is much better documented than its very early history. Accessible summaries of Turkish history in this period include Halil Inalcik, *The Ottoman Empire: The Classical Age 1300–1600*.

The tulip in Ottoman history to 1453 Yildiz Demiriz, 'Tulips in Ottoman Turkish culture and art', in Roding and Theunissen pp.57–75.

The story of Hasan Efendi Demiriz p.57.

Babur and the Turkish gardening tradition Alexander Pallis, *In the Days of the Janissaries* p.198.

The tulip as a religious symbol The Turks were not the only people to regard the flower as a religious symbol. Among the 'Pennsylvania Dutch' – German immigrants who travelled to the east coast of America from the seventeenth century – stylized three-petal tulips were used as a motif which symbolized the Holy Trinity. They were often used to adorn important papers such as birth certificates. Schloredt p.43.

2: WITHIN THE ABODE OF BLISS

Horticulture is hardly central to the history of the Ottoman Empire, and features scarcely at all in conventional histories. The best guides to the story of the tulip's time in Turkey have been accounts of Istanbul. The finest of these is certainly Philip Mansel, *Constantinople: City of the World's Desire, 1453–1924*. For the Ottoman palaces, the indispensable source is Barnette Miller, *Beyond the Sublime Porte: The Grand Seraglio of Stambul*. Dr Miller was probably the first Westerner to gain access to the inner courtyards of the Topkapi, and she did so at a time, early in the twentieth century, when they still looked much as they did in earlier days. She worked hard to reconstruct those institutions – such as the harem and the gardens – which had fallen into disuse or disrepair, and her work has formed the basis for all subsequent descriptions of Ottoman palace life.

Battle of Kosovo Noel Malcolm, *Kosovo: A Short History* pp.58–80. For the chronicler, see Anna Pavord, *The Tulip* p.31.

Bayezid Halil Inalcik, *The Ottoman Empire: The Classical Age 1300–1600* pp.14–18; John Julius Norwich, *Byzantium: The Decline and Fall* pp.343–5, 364–9.

Bayezid's shirt There is some dispute about the age of this garment. The Museum of Turkish and Islamic Arts dates it to c.1400, but Yildiz Demiriz, 'Tulips in Ottoman culture and art', in Michiel Roding and Hans Theunissen, *The Tulip: A Symbol of Two Nations* p.71, suggests that the style dates the shirt to about 1550. The tradition therefore remains unproven – but even if Demiriz is right, it is certainly not impossible that Bayezid wore a similar shirt.

Constantinople and Sultan Mehmed Mansel, chapter 1.

Sultan Mehmed's gardens Andrew Wheatcroft, *The Ottomans: Dissolving Images* pp.26–9, Mansel pp.57–8.

Sultan Suleyman and the Istanbul tulips Arthur Baker, 'The cult of the tulip in Turkey' p.240; Baytop, in Roding and Theunissen pp.52–3; Demiriz pp.57–8, 74–5. Some authorities argue that Istanbul tulips were not in fact bred until the second half of the seventeenth century (cf. Pavord pp.39, 45); the matter is unclear.

Florists in Istanbul Baytop p.51.

Sultan Selim and bulbs from Persia and Syria Ibid p.53; Baker pp.238–40.

The sultan's palace and gardens Demiriz pp.59, 67; Mansel pp.60–1, 71, 73–5, 221–2; Miller pp.4–21, 151–6; N. M. Penzer, *The Harem* pp.40, 252–60; Lavender Cassels, *The Struggle for the Ottoman Empire, 1717–1740* pp.53–4, 57–8.

The bostancis Mansel pp.74–5, 221–2; Cassels p.53; Penzer pp.62, 185.

The head gardener's race It does not seem to be known when exactly this weird custom originated. Miller pp.145, 250, n31.

3: STRANGER FROM THE EAST

The early history of the tulip in Europe – in so far as it is known or can be guessed – was first thoroughly documented by Hermann, Grafen zu Solms-Laubach, in *Weizen und Tulpe und deren Geschichte*, and summarized in English by Sir Daniel Hall, *The Book of the Tulip*. More recent research is very briefly summarized by Sam Segal, *Tulips Portrayed: The Tulip Trade in Holland in the Seventeenth Century*.

Lopo Vaz de Sampayo Vaz's connection with the tulip is also mentioned by Wilfrid Blunt in *Tulipomania* p.8n. Details of his career have been drawn from R. S. Whiteway, *The Rise of Portuguese Power in India 1497–1550* pp.208–13, 221–3. Nunho da Cunha, incidentally, was the son of Tristao da Cunha, who gave his name to the flyspeck island in the Atlantic which still forms one of the remoter outposts of the British Empire.

Monstereul Charles de la Chesnée Monstereul's book was the earliest to be entirely devoted to the tulip, and therefore carries some weight among historians of the flower.

Duration of voyages to Portugal Whiteway p.46.

Tulip hailed as something new Hall p.36.

Evidence for tulips in Europe before the sixteenth century Hall pp.17, 36–7.

Busbecq Baytop, in Roding and Theunissen, p.52; Z. R. M. W. von Martels, *Augerius Gislenius Busbequius* pp.152, 440–52. On the proper dating of Busbecq's first encounter with the tulip, see von Martels pp.449–50.

George Sandys Cited in Anna Pavord, *The Tulip* pp.35–6.

Busbecq's letters The book was *Legationis Turcicæ epistolæ quatuor* (Antwerp, 1581), and it was a best-seller in its time.

Busbecq and the introduction of the tulip Another good reason for doubting that the ambassador was personally responsible for bringing the tulip to Europe is that Busbecq frequently boasted that he had been the first to introduce the sweet flag to the West. Given the fame which the tulip had already attained by the time of his death in 1591, it seems inconceivable he would not also have claimed credit for that discovery, if he knew he had been the first to make it. Von Martels pp.450–2.

The word 'tulip' in English According to Hall, p.17, it first appeared in Lyte's translation of *Florum et Coronarium Odoratumque Nonnularum*, by Clusius's friend Rembert Dodoens, originally published in Antwerp in 1568.

Garret and Gerard Wilfrid Blunt pp.10–11; Pavord pp.104–5.

Conrad Gesner Hall p.39; Sam Segal p.3; E. H. Krelage, *Bloemenspeculatie in Nederland* pp.15–16; Hans Fischer, *Conrad Gesner 1516–1565*; for the frog story, see Jan Bondeson, 'The bosom serpent', in *A Cabinet of Medical Curiosities*. *Catalogus plantarum*, incidentally, was not published until two centuries after Gesner's death; his description of the tulip first appeared in an appendix he added to a book written by his friend Valerius Gordus, which was published in 1561.

'In the month of April ...' Quoted in Hall p.39.

Tulipa turcarum Although a species of tulip named in Gesner's honour was long thought to be that discovered at Augsburg, it would appear, according to W. S. Murray, 'The introduction of the tulip, and the tulipomania', p.19, that the species in Herwart's garden was probably *T. suavenolens* and not *T. gesneriana* at all.

Tulip seen in Italy by Johann Kentmann Segal pp. 3, 21 n6. Kentmann labelled this flower *T. Turcica*, but it appears to have been an example of the species *T. silvestris*.

The Fugger gardens R. Ehrenberg *Grosse Vermögen* p.38. See also G. Freiherr von Polnitz, *Die Fugger*. Anton Fugger, the son of the founder of the Fugger empire, offered employment to both Gesner and Clusius; the latter, despite his religious scruples (since the Fuggers bankrolled much of the Counter-Reformation), accepted.

Early tulips in England and Europe Hall p.40; Joseph Jacob, *Tulips* p.3; Blunt pp.10–11.

4: CLUSIUS

Easily the most comprehensive biography of Clusius is that published by F. W. T. Hunger in the two volumes of *Charles de L'Escluse (Carolus Clusius), Nederlandsche Kruidkundige 1526–1609*, from which work much of the material in this chapter is drawn. A popular biography by Johan Theunisz, *Carolus Clusius: Het Merkwaardige Leven van een Pionier der Wetenschap* adds a few details, mainly to elaborate on the botanist's early life. Clusius's scattered work on the tulip – which, it has to be stressed, was never remotely central to his botanical *corpus* as a whole – has fortunately been summarized, in English, by W. van Dijk, *A Treatise on Tulips by Carolus Clusius of Arras*.

Anecdote of the Flemish merchant This story was originally recorded by Clusius himself, and is mentioned by van Dijk p.8.

Thus it was in the spring of 1563 ... This part of the account is speculation on my part, but it does strike me as unlikely, if the merchant thought the tulip bulbs were onions, that anyone would have realized what they really were until they had flowered.

Execution of an uncle This was Mathieu de L'Escluse, who was actually burned in April 1567 during the Duke of Alva's attempts to put down Protestantism in the Habsburg Netherlands. Hunger I, p.97.

Extent of Clusius's correspondence The estimate of 4000 letters is based on a calculation by Hunger I, pp.98.

Clusius on the tulip Clusius first mentioned the flower in an appendix to his book on the flora of Spain, *Historia stirpium per Hispanias observatorum*, published in 1576 (pp.510–15), even though the tulip was not native to that country, and this does perhaps suggest that it was while he was travelling in Spain that he first heard about it from Rye. He elaborated considerably on its botany in a work on the flora of Austria, *Historia stirpium Pannoniæ*, published in 1583 (pp.145–69) and again in his masterpiece, *Rariorum Plantarum Historia* of 1601 (pp.137–52).

Experiments at Frankfurt This was in 1593. W. S. Murray, 'The introduction of the tulip, and the tulipomania' p.19.

Clusius's character and disposition Hunger I, 323.

Marie de Brimeu's compliment Ibid, II, 217.

Clusius's poverty Ibid. I, 111, 122.

Plant trade between the Ottomans and Vienna Theunisz p.68.

Clusius and Busbecq Clusius had already, in 1569, written to von Krafftheim asking him to obtain samples of plants from Busbecq. Hunger I, 108, 139.

Busbecq's seed van Dijk p.32.

Flower thieves Hunger I, 158; II, 115, 135; Theunisz pp.50, 78.

'... lost all his teeth' Hunger I, 180, 240.

5: LEIDEN

The biographies of Hunger and Theunisz are again the principal sources for Clusius's career at Leiden. On the university at Leiden, the course of the Dutch revolt and the historical background to the mania period, see Jonathan Israel's magisterial *The Dutch Republic: Its Rise, Greatness and Fall, 1477–1806*. The university, and particularly its famous anatomy school, was frequently mentioned by foreign visitors, and the accounts of Sir William Brereton, *Travels in Holland, the United Provinces etc ... 1634–1635* and John Evelyn, *The Diary of John Evelyn*, II make interesting reading. In discussing the tulip's botany, I have drawn on Daniel Hall, *The Book of the Tulip* and E. van Slogteren, 'Broken tulips' in *The Daffodil and Tulip Yearbook*.

Clusius in Frankfurt Hunger II pp.153–4, 164–5, 167, 172–5.

Arrival in Leiden Hunger I pp.210–13.

Leiden Israel pp.308, 328; Paul Zumthor, *Daily Life in Rembrandt's Holland* pp.8, 12, 23, 239.

Dutch Revolt Israel pp.169–75, 181–2.

University of Leiden Israel pp.569–72, Schama pp.57, 175; Brereton pp.41–2; Evelyn pp.51–4; Zumthor p.154.

The Leiden hortus Hunger I pp.189–94, 214–18; Hunger II p.4; Israel pp.571–2, 1043; Brereton p.42.

'True monarch of the flowers' From a letter dated 28 February 1602, quoted in Hunger I p.269.

Walich Ziwertsz. Nicolaes van Wassenaer, *Historisch Verhael* IX, section April–October 1625, p.10; A. Hensen, 'De vereering van St Nicolaas te Amsterdam in den Roomschen tijd' in *Bijdragon voor de Geschiedenis van het Bisdom Haarlem*, 43 (Haarlem 1925) p.187.

Clusius on tulips W. van Dijk, *A Treatise on Tulips by Carolus Clusius of Arras* pp.7–32.

Botany of the tulip Sam Segal, *Tulips Portrayed* pp.5–12; Hall, pp.99–110; W. S. Murray, 'The introduction of the tulip, and the tulipomania' pp.21–3.

Offsets John Mather, *Economic Production of Tulips and Daffodils* p.44.

Rosen, Violetten and Bizarden tulips E. H. Krelage, *Bloemenspeculatie in Nederland* p.33, makes the point that these category names were only introduced in the nineteenth century, but as they are so convenient we will use them here. The *Violetten* varieties, incidentally, are also sometimes known as *bybloemen* tulips.

'Superbly fine' and 'rude' Ibid p.21.

Attempts to replicate breaking Anna Pavord, *The Tulip* p.11.

Solution to the problem of breaking Hall pp.104–6.

Clusius and the demand for tulip bulbs Hunger I pp.214, 237.

Theft of bulbs Theunisz p.120; Hunger I pp.237–8, 241; Hunger II p.197.

'The seventeen provinces were amply stocked' Cited in Wilfrid Blunt, *Tulipomania* p.9.

6: AN ADORNMENT TO THE CLEAVAGE

The early history of the tulip in the United Provinces and France is not especially well documented. The basic details given here are summarized from Krelage's books and from the works of contemporary gardeners such as Abraham Munting, *Waare Oeffening der Planten*, from W. S. Murray, 'The introduction of the tulip, and the tulipomania', and from Sam Segal, *Tulips Portrayed*; the last-named also includes a useful discussion of what is known about seventeeth-century tulip books.

The rose as empress of the garden Paul Zumthor, *Daily Life in Rembrandt's Holland* p.49.

Monstereul's eulogy Cited by Segal p.4.

Lobelius The Latinized name of Mathias de l'Obel, whose work on tulips was published in a French herbal of 1581. Segal p.3.

Varieties of tulip Segal p.4; Murray p.21. These totals exclude Turkish species, which by the eighteenth century numbered over 1300 by themselves.

Early tulip lovers E. H. Krelage, *Bloemenspeculatie in Nederland* pp.23–4; Krelage, *Drie Eeuwen Bloembollenexport* pp.6, 17.

The tulip in France Krelage, *Bloemenspeculatie* p.29; Abraham Munting, *Naauwkeurige Beschryving der Aardgewassen* pp.907–11; Peter Garber, 'Tulipmania' p.543. Although dealt with by contemporary garden writers, the history of this early French tulip mania is still obscure and would probably repay some original research.

The tulip connoisseurs Chrispijn van de Passe, *Een Cort Verhael van den Tulipanen ende haere Oefeninghe* ..., contemporary pamphlet, np, nd (c.1620?) in Stadsbibliotheek, Haarlem, p.4; Krelage, *Drie Eeuwen Bloembollenexport* p.6.

Paulus van Beresteyn E. A. van Beresteyn and W. F. del Campo Hartman, *Genealogie van het Geslacht van Beresteyn* p.134.

Jacques de Gheyn L. Q. van Regteren Altena, *Jacques de Gheyn: Three Generations* vol.I pp.2–3, 14, 38, 40, 59, 66, 69–70, 131–2, 153.

Guillelmo Bartolotti G. Leonhardt, *Het Huis Bartolotti en zijn Bewoners* pp.14–15, 39–40; Israel p.348.

The Golden Age J. L. Price, *Culture and Society in the Dutch Republic During the 17th Century*; Jonathan Israel, *The Dutch Republic* pp.547–91.

The Lord Offerbeake's garden William Brereton, *Travels in Holland, the United Provinces etc* ... *1634-1635* pp.44–5.

Dutch country houses Simon Schama, *The Embarrassment of Riches* pp.292–5; Krelage, *Bloemenspeculatie* pp.7, 27–8.

Jokes in church Geoffrey Cotterell, *Amsterdam: The Life of a City* p.119. The usual fine was 6 stuivers per joke.

Jacob Cats Schama pp.211, 293, 437.

Of de Moufe-schans Petrus Hondius, *Dapes Inemptae, of de Moufe-schans*. On the true ownership of the Moufe-schans, which is sometimes incorrectly said to have been Hondius's own home, see *Nieuw Nederlandsch Biographisch Woordenboek*, VIII pp.812–13.

'All these fools want ...' English translation from Segal p.16.

The Prince of Orange's garden Brereton, pp.34–5.

7: THE TULIP IN THE MIRROR

My discussion of Semper Augustus is based, as all such discussions must be, on the chronicle of Nicolaes Jansz. van Wassenaer. Van Wassenaer, the son of an Amsterdam physician, taught at the Latin School in Haarlem and then in Amsterdam before becoming a professional writer (and part-time physician) after 1612. His chronicle, *Historisch Verhael aller Gedencwaerdiger Gheschiedenissen*, which is in general one of the most reliable available, is the principal source of information on the flower.

The passages on the progress of the tulip craze are based as before on the works of E. H. Krelage, supplemented by those of Nicolaas Posthumus, 'Die speculatie in tulpen in de jaren 1636 en 1637' parts 1–3, and 'The tulip mania in Holland in the years 1636 and 1637', in W. C. Scoville and J. C. LaForce (eds), *The Economic Development of Western Europe* and that of Peter Garber, 'Tulipmania' pp.535–60. Information on Dutch gardens of the period is drawn from Paul Zumthor, *Daily Life in Rembrandt's Holland* and Simon Schama's *Embarrassment of Riches*.

Information on tulip books comes from Sam Segal, *Tulips Portrayed*, and E. H. Krelage, *Bloemenspeculatie in Nederland*. The *Hortus Floridus* of Chrispijn van de Passe has been the subject of some research; see Spencer Savage, 'The "Hortus Floridus" of Crispijn vande Pas' pp.181–206 and Eleanour Rohde, *Crispian Passeus's 'Hortus Floridus'*. Savage's English translation appeared in the 1970s: *Hortus Floridus: The Four Books of Spring, Summer, Autumn and Winter Flowers, Engraved by Crispin van de Pas.*

Adriaen Pauw Jonathan Israel, *The Dutch Republic* pp.159, 319, 458–9, 518–19, 522–33; H. W. J. de Boer, H. Bruch et al, *Adriaan Pauw (1585–1653): Staatsman en Ambachtsheer* pp.20–7. Today only a small portion of the Heemstede estate can still be seen; the rest has been swallowed up by Haarlem and now forms one of the southernmost suburbs of the city.

Pauw's mirrored garden Van Wassenaer V, p.40 and verso. It is possible that the *Violetten* variety Pauw, mentioned by Krelage, *Bloemenspeculatie in Nederland* p.138, was created by him or at least named for him.

Semper Augustus Van Wassenaer V, p.40 verso and 41; p.111 and verso; IX p.10; Krelage, *Bloemenspeculatie* pp.32–3, 68; Garber p.537; Segal pp.8–9.

The ownership of Semper Augustus In recent years, several authorities have confidently stated that the owner of Semper Augustus was none other than Adriaen Pauw, but that is because they have not read van Wassenaer's work carefully. In fact, although the chronicler did see specimens of the flower, and did visit the garden at Heemstede, nowhere does he link the two, and the description he gives of Pauw's single tulip bed makes it unlikely that Semper Augustus – a flower any connoisseur would have planted in solitary splendour – would have been grown there.

There are several unreferenced anecdotes which suggest other Semper Augustus bulbs were sold, but until they can be confirmed in contemporary records I would be reluctant to accept them at face value. Krelage p.65n says that an Amsterdammer sold a Haarlemmer the flower on condition neither sold any further Semper Augustus bulbs without notifying the other first. The Amsterdam connoisseur later succumbed to the temptation of 3000 guilders and a cabinet worth 10,000 guilders for a single bulb. When the Haarlemmer discovered this deception, he in turn sold three bulbs for 30,000 guilders. Similarly, Munting, writing some thirty-five years after the mania, quoted an unreferenced bookkeeper's entry which reads: 'Sold to N.N., a Semper

Augustus weighing 123 azen, for the sum of 4600 florins.* Above this sum a
new and well-made carriage and two dapple grey horses with all accessories to
be delivered within two weeks, the money to be paid immediately.' He also
alleges that a bulb was sold for 5500 florins at public auction. See Abraham
Munting, *Naauwkeurige Beschryving der Aardgewassen* pp.907–11.

Balthasar and Daniël de Neufville D. M. van Gelder de Neufville, 'De oudste
generatics van het geslacht de Neufville' in *De Nederlandsche Leeuw* pp.6–8;
Krelage pp.129, 140. These varieties bore the corrupted name 'de Novil'.

Tulip growers F. W. T. Hunger, *Charles d'Escluse (Carolus Clusius)* I, p.241; II,
p.251.

Henrik Pottebacker Segal p.8; Krelage pp.127, 138.

Rhizotomi and apothecaries Hunger I, p.303–06; Krelage, *Drie Eeuwen Bloembollenexport* p.17. On the unreliability of apothecaries, see Zumthor pp. 73,
157.

The tulip as aphrodisiac Sam Segal and Michiel Roding, *De Tulp en de Kunst.
Verhaal van een Symbool* p.22. The contemporary English garden writer John
Parkinson mentions the supposed aphrodisiac qualities of the flower in *Paradisus
terrestris* (1629), confessing however: 'For force of Venereous quality, I cannot
say ... not having eaten many.' Quoted in Wilfrid Blunt, *Tulipomania* pp.10–11.

Actors in the early tulip trade Nicolaas Posthumus, 'De speculatie in tulpen'
(1927) pp.11–15.

Gardens outside Haarlem J. J. Temminck et al, *Haarlemmerhout 400 Jaar. 'Mooier
is de Wereld Nergens.'* pp.98–9.

Tulip nomenclature Krelage pp.33–7, 128.

'If a change in a Tulip is effected ...' Cited by W. S. Murray, 'The introduction
of the tulip, and the tulipomania' p.24.

Pieter Bol and Barent Cardoes Krelage p.42; Schama p.356. Cardoes died late
in 1657 (Haarlem Burial Registers 72, fol.100), but the business he established
was still in existence in the eighteenth century.

Francisco da Costa Unsurprisingly, da Costa's business was a very sound one,
and it survived the mania and continued until at least 1645. Krelage pp.42–3,
55; Krelage, 'Het manuscript over den tulpenwindhandel uit de verzameling
Meulman' p.30.

Bulb exports Today, fully two-thirds of Dutch bulbs are exported and the
largest single producer, Germaco, ships some 35 million bulbs a year overseas –
many of them to British supermarket chains.

Emanuel Sweerts Krelage p.25.

Tulip books The earliest known flower book dates from 1603 and is French.
Books portraying only tulips came into existence as the mania developed; the
oldest of these dates from about 1635. Segal and Roding pp.78–81; Segal *Tulips
Portrayed* pp.17–20; Paul Taylor, *Dutch Flower Painting 1600–1720* pp.10–12.

---

* A florin had the same value as a guilder.

Van Swanenburch's tulip book This book is now in the archives of the Nederlandsche Economisch-Historisch Archief in Amsterdam. The notes on prices appear to have been written by the book's – anonymous – original owner.

Cos's tulip book This manuscript, correctly titled *Verzameling van een Meenigte Tulipaanen ...*, was written in 1637. (Oddly there do not seem to be any other records of a florist called Cos in the city archives, although Krelage does note the existence of a tulip variety named Kos.) It is now in the Universiteitsbibliotheek at Wageningen.

Travelling bulb sellers Anna Pavord, *The Tulip* p.53.

8: FLORISTS

The social history of the United Provinces during the Golden Age is ably dealt with by A. T. van Deursen, *Plain Lives in a Golden Age: Popular Culture, Religion and Society in Seventeenth Century Holland.* Details of day-to-day life are added by Paul Zumthor, *Daily Life in Rembrandt's Holland.* Among contemporary authors, the greatest authority was generally reckoned to be Sir William Temple, whose *Observations Upon the United Provinces of the Netherlands* did not, unfortunately, appear until 1673, well after the mania. This short book was, nevertheless, based on the author's observations during visits dating back to 1652, and as Temple was for some time the English ambassador to the United Provinces, and took a keen professional interest in the reasons for Dutch success, his work is far more thoughtful and analytical than the muddled impressions of travellers, as well as being considerably less superficial.

Physical description of the United Provinces Temple pp.95, 113–14; Zumthor p.277; Jonathan Israel, *The Dutch Republic* pp.1–3, 9–14.

'An universall quagmire ...' The Englishman was the propagandist Owen Felltham, and his work was published when Anglo-Dutch antagonism reached its peak in the middle of the seventeenth century. Felltham's views of the Dutch need to be seen in this context. Cited by Simon Schama, *The Embarrassment of Riches* p.44.

The English ambassador ... Temple pp.95, 113–14.

The classes of Dutch society Israel pp.330, 337–53, 630–8; van Deursen pp.4–8, 13, 32, 47–8, 194; Zumthor pp.232–41; Schama pp.19–21, 316, 579–81.

Guilds and the working day Van Deursen pp.5, 11; Zumthor pp.5–6, 53.

Food Van Deursen pp.4, 19–20, 82; Schama pp.162–4, 169–70, 230; Zumthor pp.67–74; Geoffrey Cotterell, *Amsterdam: The Life of a City* pp.24, 48; William Brereton *Travels in Holland, the United Provinces etc ... 1634–1635* p.6.

Cleanliness Van Deursen pp.19, 41; Zumthor pp.137–9, 169; Brereton p.68.

Population Israel p.328.

Baudartius and the pressure of overpopulation Van Deursen pp.3–4, 8.

Spread of the fashion for gardening in the Netherlands Cotterell pp.88, 131; Brereton p.38; Peter Mundy, *The Travels of Peter Mundy* vol.IV p.75; Segal p.8; Joseph Bulgatz, *Ponzi Schemes* p.86.

Dutch savings Temple p.102.

The gambling impulse Van Deursen pp.67–8, 105–6; Schama pp.306–7, 347; Zumthor p.76.

9: BOOM

The course of the mania is set out best in E. H. Krelage's *Bloemenspeculatie in Nederland.* A general summary of events, with rather more interpretation, can be found in Nicolaas Posthumus, 'The tulip mania in Holland in the years 1636 and 1637', in W. C. Scoville and J. C. LaForce (eds), *The Economic Development of Western Europe* pp.138–49.

Hoorn Jonathan Israel, *The Dutch Republic* pp.317–18.

The tulip house A. van Damme, *Aanteekeningen Betreffende de Geschiedenis der Bloembollen* pp.23–4. According to van Damme, the house was renovated in 1755, at which time the stone tulips were inscribed with some memorial to the mania. At some time in the 1880s or early 1890s, the house was demolished, and the tulips were purchased by J. H. Krelage, one of the leading tulip growers of Haarlem and the father of E. H. Krelage, the tulip historian, and set in the wall of his library. Van Damme, incidentally, describes the chronicle from which he drew many of his details as Velius's, but in fact Velius's work runs no further than 1630. He must therefore have meant a continuation of the original chronicle. The reliability of this work is not entirely clear. From the context in which the chronicler mentions the tulip house, it seems the passage may not be contemporary.

The development of the tulip mania pp.140–2; Krelage pp.42, 49–52.

'A contemporary chronicler ...' Lieuwe van Aitzema, *Saken van Staet en Oorlogh* p.504. Like many of the prices cited by historians of the mania, van Aitzema's seem to be drawn from the fictionalized *Samenspraecken,* three pamphlets published in 1637 which purported to record conversations between a tulip dealer and his friend. See below and chapter 11 for details.

Generael der Generaelen van Gouda Krelage pp.35, 49. Schama says that Gouda was one of the cheapest and least spectacular varieties, which is not correct.

Later prices quoted for Semper Augustus Krelage pp.32–3, 68; Garber, 'Tulipmania' p.537; Sam Segal, *Tulips Portrayed* pp.8–9.

Soap Cf. Israel p.347.

Land in Schermer polder; the merchant lover Krelage p.30, citing one of the pamphlets published during the mania.

Anecdotes of a sailor and an English traveller The story of the sailor is recorded by J. B. Schuppius as a memory of his youth in Holland, according to Hermann, Grafen zu Solms-Laubach, *Weizen und Tulpe und deren Geschichte* p.76. It was famously retold in considerably embellished form by Charles Mackay, *Memoirs of Extraordinary Popular Delusions and the Madness of Crowds* p.92. Mackay tells the – unreferenced – story of the Englishman on the next page. Peter Garber has drawn attention to the fundamental implausibility of these anecdotes in 'Tulipmania' p.537&n.

Dutch recession Israel pp.314–15.

Weavers Those who note the predominance of linen workers among the tulip maniacs include Posthumus, p.143.

Sales by bulb and by the bed Ibid p.141.

Trades of Jan Brants and Andries Mahieu Posthumus, 'De speculatie in tulpen in de jaren 1636 en 1637' part 2 (1927) pp.13–14.

Sales between April and August All the early records of tulip trading are dated between April and August. Posthumus, ibid, pp.11–15; Posthumus, 'The tulip mania in Holland' p.141.

The windhandel Schama pp.358–9.

The futures trade Marjolein 't Hart, Joost Jonker and Jan Luiten van Zanden (eds) *A Financial History of the Netherlands* pp.53–4; Schama pp.339, 349–50; Jan de Vries and Ad van der Woude, *The First Modern Economy: Success, Failure and Perseverance of the Dutch Economy 1500–1815* p.151; Schama pp.338–9; Paul Zumthor, *Daily Life in Rembrandt's Holland* p.262.

Bans on futures trading 't Hart et al p.55.

Trading by the ace Krelage pp.46–8.

Gerrit Bosch Alkmaar notarial archive vol.113 fol.71vo–72vo, 23 July 1637 (copy in the Posthumus Collection, Netherlands Economic History Archive).

Profit on spice voyages Israel p.320.

David de Mildt Posthumus, 'De speculatie in tulpen' (1927) p.16.

Henrick Lucasz. and Joost van Haverbeeck Ibid pp.19–20.

Jan Admirael Ibid pp.17–18, 21–2.

The value of a bulb The best data come from the auction held at Alkmaar in February 1637, where several bulbs of the same variety, but of different weights, were sold to the same bidders in the course of a single day. See van Damme pp.92–3.

Tulip companies Posthumus, 'De speculatie in tulpen' (1927) pp.26, 32–6.

Bulbs per ace and per thousand aces Cf. van Damme pp.92–3.

Bulbs bought to plant and trade Posthumus, 'De speculatie in tulpen' (1927) pp.24–5.

'They came from all walks of life ...' Ibid (1926) pp.3–99.

The Samenspraecken These three important pamphlets were reprinted by Posthumus in the *Economisch-historisch jaarboek* (1926), pp.20–99. They have been discussed by Krelage, *Bloemenspeculatie in Nederland* pp.70–3 and by the same author in *De Pamfletten van den Tulpenwindhandel 1636–1637* pp.2–4; and also by W. S. Murray, 'The introduction of the tulip, and the tulipomania', pp.25–7, Joseph Jacob, *Tulips* pp.10–12, Sam Segal, *Tulips Portrayed* pp.13–15, Zbigniew Herbert, *Still Life with a Bridle* pp.57–8, and Schama pp.359–60. None of these accounts, incidentally, agrees with any of the others on precisely how the information in the *Samenspraecken* should be interpreted, a mute testimony to the remarkable obscurity of the text of the original pamphlets.

Payments in kind As noted, these examples, too, derive from the *Samenspraecken*. Cf. Joseph Bulgatz, *Ponzi Schemes* p.97.

Aert Ducens Posthumus, 'De speculatie in tulpen' (1927) p.38. In 1643, van de Heuvel's wife appeared before a notary and confirmed that this agreement had been cancelled after the tulip market crashed.

Jeuriaen Jansz. Posthumus, ibid pp.27–8. In this case the seller's name is given as 'Cresser', but the records of the mania are full of misspelled surnames and it is almost certainly Creitser who is meant.

Cornelis Guldewagen Ibid, pp.61–5, 72–4.

Abraham de Goyer Ibid (1934) pp.231–2.

'Null and void ...' Ibid (1927) p.85.

Cases of deceit and fraud Segal p.12; W. S. Murray p.25.

'Everything that could be called a tulip ...' van Aitzema p.504.

10: AT THE SIGN OF THE GOLDEN GRAPE

My account of Dutch tavern life has been pieced together from numerous secondary sources, the most significant being those of A. T. van Deursen and Simon Schama. The English travellers Fynes Moryson, William Brereton and Peter Mundy all make some mention of the subject, and their personal experiences add colour to the general remarks of the social historians. Haarlem's brewing industry is described in S. Slive (ed), *Frans Hals* (The Hague: SDU, 1990). The taverns of Haarlem are touched on by S. Groenveld, E. K. Grootes, J. J. Temminck et al, *Deugd Boven Geweld. Een Geschiedenis van Haarlem 1245–1995*, which is more rewarding than an English translation of its title ('Virtue Above Violence') might suggest; and the brothels of the Haarlemmerhout are rather tentatively passed over by J. J. Temminck et al in the even less enticingly titled *Haarlemmerhout 400 Jaar. 'Mooier is de Wereld Nergens'* ('400 Years of Haarlem Wood: Nowhere in the World Is More Beautiful'). Thankfully Geoffrey Cotterell's anecdotal history *Amsterdam: The Life of a City* adds some more entertaining details about the role that food and drink played in Dutch life.

The Amsterdam stock exchange Marjolein 't Hart, Joost Jonker and Jan Luiten van Zanden (eds), *A Financial History of the Netherlands* pp.53–6; Cotterell pp.85–6; Schama pp.348–50; Brereton pp.55–6.

De la Vega on small-time traders Cited by Schama p.349. The descriptions of traders' behaviour date to somewhat after the mania period – to the 1680s to be exact – and it may not necessarily have been so exaggerated in the 1630s.

Ubiquity of inns Van Deursen pp.101–2.

Pub names Schama p.202; Zbigniew Herbert, *Still Life with a Bridle* p.58.

Prostitution Van Deursen pp.97–100.

'Impudent whores' Brereton p.55. He was referring to the prostitutes of Amsterdam.

Beginnings of the tavern trade Posthumus 'De speculatie in tulpen' (1927) p.19.

Taverns involved in the tulip mania Haarlem inns definitely known to have been involved in the mania include Van de Sijde Specxs ('The Flitch of Bacon'), De Vergulden Kettingh ('The Gilded Necklace'), 't Oude Haentgen ('The Little Old Hen'), the Toelast in the Grote Markt and De Coninck van Vranckrijck ('The King of France'). In Amsterdam, De Mennoniste Bruyloft ('The Mennonite Wedding') also served as a centre of tulip dealing. Posthumus, 'De speculatie in tulpen' (1927) pp.24, 42–3, 83 and (1934) p.233.

The Quaeckels Cornelis Quaeckel senior was born around 1565 and married, in 1587, Trijn or Catharina Cornelisdr. Duyck. From 1609 he ran a tavern called the Bellært in the Kruisstraat in Haarlem, but he also grew crops and tulips on an allotment near the Janspoort and on land he rented from the Lord of Brederode near the castle of Huis ter Kleef. Roads leading to both locations were named Quaeckelslaan after the family. There seems to be no record that Quaeckel's eldest son, Cornelis Cornelisz., had any involvement in the tulip trade, but he did testify in favour of the allegedly heretical painter Torrentius during his persecution in 1627. Cornelis Cornelisz. was Haarlem's collector of taxes on soap until 1626 and lived until at least 1650. Jan Quaeckel, his tulip-trading brother, was born in 1601–2 and buried in Haarlem on 10 November 1661. G. H. Kurtz, 'Twee oude patriciërshuizen in de Kruisstraat' in *Jaarboek Haerlem*, 1961, p.20; Haarlem Municipal Archives, Notarial Records vol.129 fols.72; vol.123vo; vol.139 fol.27vo–28; vol.149, fol.210; vol.150 fols.273–273vo, 394vo; Haarlem burial registers vol.73 fol.100vo. E. H. Krelage, *Bloemenspeculatie in Nederland* pp.134–6, gives details of the tulip species created by Cornelis Quaeckel senior.

Haarlem Groenveld et al, pp.144, 172–4, 177.

Street lighting Lighting – using hundreds of lamps burning vegetable oil – was eventually introduced in Amsterdam in 1670, with such success that it quickly spread to other Dutch cities and then across Europe. Jonathan Israel, *The Dutch Republic* p.681.

Peat fires Mundy, *The Travels of Peter Mundy* pp.64–5; Monsieur de Blainville, *Travels Through Holland* ... (London, 1743) I, 44.

Smoking Schama pp.194–8; van Deursen pp.103–4.

Weapons Van Deursen pp.110–11. A ban on weapons was instituted by the States of Holland in 1589, backed up in many cases by local legislation.

Paintings John Stoye, *English Travellers Abroad, 1604–1667* p.294, records comments about the magnificence of the paintings to be found in Dutch taverns by the English travellers Sir Dudley Carleton (1616) and Robert Bargrave (1656).

Drunkenness and drink Ibid p.162; Cotterell p.73; Brereton pp.11–12.

Quantity of beer consumed in Haarlem Paul Zumthor, *Daily Life in Rembrandt's Holland* p.72, citing J. van Loenen, *De Haarlemse Brouwindustrie voor 1600* (Amsterdam, 1950) p.53.

Cost of an evening's drinking Fynes Moryson, travelling in 1592, paid between 12 and 20 stuivers for a meal, complaining that this high price was the result of his paying for the ale consumed by his travelling companions, who spent the evening roistering by the fire. Moryson, *An Itinerary* pp.89–90.

Consumption of alcohol Zumthor p.175; Schama pp.191, 199.

Théophile de Viau Cited in Zumthor p.173.

Number of breweries Groenveld et al. p.176; H. L. Janssen van Raaij, *Kroniek der Stad Haarlem van de Vermoedelijke Stichting der Stad tot het Einde van het Jaar 1890*, entry for 1628.

The tavern trade Posthumus, 'De speculatie in tulpen (1926) pp.20–99; Zumthor p.175.

Wine Zumthor pp.173–4.

11: THE ORPHANS OF WOUTER WINKEL

The little that we know about Wouter Winkel and his family is contained in documents from the Stad Archief at Alkmaar. These were recovered and published by A. van Damme among a collection of solicitors' acts and pamphlets concerning the mania which appeared in a series of articles published in a bulb growers' periodical around the turn of the nineteenth century. Van Damme's articles were subsequently collected and republished in book form in *Aanteekeningen Betreffende de Geschiedenis der Bloembollen: Haarlem 1899–1903*. Van Damme's archival work, along with that of Posthumus, provides the bedrock of all serious studies of the tulip mania, including those of E. H. Krelage, and has not yet been supplemented in any significant way by more modern research.

Wouter Winkel Van Damme pp.91–3.

Alkmaar Jan de Vries, *The Dutch Rural Economy in the Golden Age, 1500–1700* pp.157–9; Paul Zumthor, *Daily Life in Rembrandt's Holland* pp.29–30, 55.

School age Simon Schama, *The Embarrassment of Riches* p.538.

Winkel's collection The surviving records indicate that Winkel was in business with one or more partners, but it would appear that the stock was divided in August 1636 and that the tulips auctioned at Alkmaar were Winkel's share of a larger collection. Van Damme p.92.

Winkel as a grower It is extremely probable, but not quite certain, that Winkel cultivated tulips. Certainly the trustees of the Alkmaar Orphans' Court did have his bulbs physically in their possession after lifting-time, and on their instructions they were later replanted. Because bulbs had to be paid for on delivery, and because it seems improbable in the extreme that a tavern-keeper could have had the thousands of guilders-worth of liquid assets required to purchase such a valuable collection, I find it difficult to believe that the trustees collected bulbs that other growers had readied for delivery to the Oude-Schutters Doelen and that Winkel simply dealt in bulbs which he purchased for delivery after lifting, and planned to sell on before autumn.

Dutch orphanages and old peoples' homes Zumthor pp.100–1.

The grower from Blokker E. H. Krelage, *De Pamfletten van den Tulpenwindhandel 1636–1637* p.30.

The quality of the bidders at Alkmaar The only bidders we actually know about were Gerrit Adriaensz. Amsterdam of Alkmaar, Jan Cornelisz. Quaeckel of Haarlem, and Pieter Gerritsz. van Welsen, all wealthy and influential growers and dealers. Posthumus, 'De speculatie in tulpen' (1927) p.81. See the next chapter for details.

The auction Van Damme pp.91–3.

'Thus Admirael Liefkens ...' Krelage, *Bloemenspeculatie in Nederland* p.49.

Hendrick Pietersz. Posthumus (1927) pp.40–1.

Van Gennep's ledger Ibid pp.39–40.

Utrecht and Groningen Representatives from Utrecht attended a conference at Amsterdam to try to control the collapse in the bulb trade (see the next chapter for details). The apothecary Henricus Munting (1583–1658), who later founded the botanical garden at the University of Groningen, dealt in bulbs in the town of Groningen during the mania period, according to his son Abraham Munting in his *Naauwkeurige Beschryving der Aardgewassen* p.911; see chapter 13. See also *Nieuw Nederlandsch Biographisch Woordenboek* vol.VI pp.1044–5.

Tulip speculation in France Munting p.911.

Numbers involved in Utrecht A list of the thirty-nine nurserymen who met in Utrecht on 7 February 1637 to elect representatives to a conference of growers due to be held in Amsterdam is given by Posthumus, 'De speculatie in tulpen' (1927) p.44.

Centres of the tulip trade Krelage pp.83–4.

Bulbs change hands 10 times in a day Ibid p.77.

Peak prices Lieuwe van Aitzema, *Saken van Staet en Oorlogh* p.504; Posthumus p.79; Krelage p.52.

10 million guilders Van Aitzema p.503.

Bank of Amsterdam Based on 1375 accounts averaging 2500 guilders apiece. Cf. Marjolein 't Hart, Joost Jonker and Jan Luiten van Zanden (eds) *A Financial History of the Netherlands* pp.46–7.

Dutch East India Company Ibid p.54.

The Black Tulip Alexandre Dumas, *The Black Tulip*; Wilfrid Blunt, *Tulipomania* p.17.

Trade in pound-goods Krelage pp.51–2.

12: BUST

The principal sources of information on the crash are the solicitors' acts of Haarlem and Amsterdam collected and published by Nicolaas Posthumus in 'De speculatie in tulpen in de jaren 1636 en 1637' parts 1–3, *Economisch-historisch jaarboek* 1926, 1927, 1934. These, however, relate almost entirely to disputes between growers and connoisseurs and need to be used with caution.

The crash E. H. Krelage, *Bloemenspeculatie in Nederland* p.80; Posthumus, 'The tulip mania in Holland in the years 1636 and 1637', in W. C. Scoville and J. C. LaForce (eds), *The Economic Development of Western Europe* pp.144–5.

Gaergoedt's plight Posthumus, 'Die speculatie in tulpen' (1926) pp.33–9.

Henricus Munting Abraham Munting, *Naauwkeurige Beschryving der Aardgewassen* p.911; *Nieuw Nederlandsche Biographisch Woordenboek* vol.VI pp.1044–5; W. S. Murray, 'The introduction of the tulip, and the tulipomania' p.29.

Geertruyt Schoudt Posthumus, 'Die speculatie in tulpen' (1927) pp.48–9.

'According to one contemporary ...' He was Abraham Munting, the son of Henricus Munting of Groningen, whose price data appear in *Naauwkeurige Beschryving der Aardgewassen* p.910.

Prices in May 1637 These examples are drawn from the *Samenspraecken* and thus probably need to be treated with a certain caution. Posthumus, 'De speculatie in tulpen' (1927) pp.80–1&n.

'Some florists did travel ...' The fictional Gaergoedt was an example of the breed. Posthumus, 'De speculatie in tulpen' (1926) p.24.

The Mennonite Wedding Ibid (1934) p.233–4.

Van Cuyk Ibid p.235.

Van Goyen Krelage pp.65–6; A. van Damme *Aanteekeningen Betreffende de Geschiedenis der Bloembollen* pp.21–2; C. Vogelaar, *Jan van Goyen* pp.13–20.

Gerrit Amsterdam Posthumus, 'De speculatie in tulpen' (1927) p.81.

Willem Lourisz Van Damme pp.94–7.

Boortens and van Welsen Posthumus, 'De speculatie in tulpen' (1927) p.53–5.

Jan Quaeckel at Alkmaar Municipal Archives, Haarlem, Notarial Registers vol.149 fol.210, 1 September 1639.

Jan Admirael Posthumus, 'De speculatie in tulpen' (1927) pp.69–70; (1934) pp.236–7.

Meeting at Utrecht Krelage p.81.

Meeting at Amsterdam Posthumus, 'De speculatie in tulpen' (1927) p.49; Krelage pp.83–4; Joseph Bulgatz, *Ponzi Schemes, Invaders from Mars* p.103.

'An ominous caveat ...' Cf. Wilfrid Blunt, *Tulipomania* p.16.

13: GODDESS OF WHORES

For Dutch tulip pamphlets, see E. H. Krelage, *De Pamfletten van den Tul-penwindhandel 1636–1637*, which reprints all known examples but the three *Samenspraecken*. (These had already been published by Posthumus in his article in the *Economisch-historisch jaarboek*, 1926.) On the various conspiracy theories of the tulip mania, see Krelage, 'Het manuscript over den tulpenwindhandel uit de verzameling Meulman'. On the liquidation, Posthumus's three-part collection of contemporary sources entitled 'De speculatie in tulpen in de jaren 1636 en 1637' is again invaluable.

Dr Tulp T. Beijer et al, *Nicolaes Tulp. Leven en Werk van een Amsterdamse Geneesheer en Magistraat* pp.15–19, 49–51; E. Griffey, 'What's in a name? Forging an identity: portraits of Nicholaes Tulp (1593–1674)' in *Dutch Crossing* 21 (1997) pp.3–43; Geoffrey Cotterell, *Amsterdam: The Life of a City* pp.125–6; Simon Schama, *The Embarrassment of Riches* pp.171, 186–7.

Adolphus Vorstius The anecdote of Vorstius the tulip-hater is recounted by several authors, although there seems to be no contemporary authority to vouch for its truth. See Wilfrid Blunt, *Tulipomania* p.15 and Zbigniew Herbert, *Still Life with a Bridle* p.60. For Vorstius himself, see William Brereton, *Travels in Holland, the United Provinces etc ... 1634–1635* pp.40–1. Vorstius's father, himself a professor at Leiden, had delivered Clusius's funeral elegy; *Nieuwe Nederlandsche Biographisch Woordenboek* vol.IV p.1411.

Kappists Joseph Bulgatz, *Ponzi Schemes, Invaders from Mars* p.99.

'A steady stream of broadsides' About forty-five examples printed between December 1636 and March 1637 are known to have survived, but given the ephemeral nature of such products, the number actually produced was almost certainly greater.

The role of pamphlets Although most of the surviving broadsides are unoriginal and contain little that is new, they are often unintentionally revealing. It is particularly instructive to compare the relatively mild tones of the early pamphlets with the increasingly bitter and sarcastic prints which began to appear when the craze was at its peak in January 1637; this adds weight to the suggestion that the tulip trade had remained fairly sober and responsible until quite late in 1636, and flared into true mania only at the end of the year for a matter of a few weeks. On pamphlets generally, see E. Craig Harline, *Pamphlets, Printing and Political Culture in the Early Dutch Republic* and Tessa Watt, *Cheap Print and Popular Piety, 1550–1640* pp.264–6.

Pamphlets commissioned by growers or connoisseurs Cf. Krelage's pamphlets nos. 9, 14, 33, 36.

Flora in the pamphlets Krelage, *Pamfletten* pp.88–91, 109–11, 149, 160, 164–7, 187–8.

The legend of Flora This retelling of the myth appeared in the first of the *Samenspraecken tusschen Waermondt ende Gaergoedt.* See Posthumus, 'De speculatie in tulpen' (1926) p.24. See also Sam Segal and Michiel Roding, *De Tulp en de Kunst* p.23 and Segal, *Tulips Portrayed* p.15.

Artistic depictions of the mania Segal pp.12–15; Schama pp.363–6; Bulgatz pp.106–7.

Resolutions of Haarlem City Council Aanteekeningen van C. J. Gonnet Betreffende de Dovestalmanege in de Grote Houtstraat, de Schouwburg op het Houtplein, het Stadhuis in de Frase Tijd, Haarlemse Plateelbakkers en Plateelbakkerijen en de Tulpomanie van 1637– 1912, Municipal Archives, Haarlem; Posthumus, 'De speculatie in tulpen' (1927) pp.51, 57; Krelage, *Bloemenspeculatie in Nederland* p.93.

Hoorn's plea to the States of Holland Posthumus, ibid p.52.

'Only two of the 44 ...' They were Burgomaster Jan de Waal and councillor Cornelis Guldewagen. Posthumus, ibid pp.61–4, 73–4; *Heerenboek* I, Municipal Archives, Haarlem.

'One anonymous pamphleteer ...' Krelage, 'Het manuscript over den tulpenwindhandel' pp.29–30.

Blame placed on bankrupts, Jews and Mennonites Ibid; A. T. van Deursen, *Plain Lives in a Golden Age* pp.32–3; Krelage, *Pamfletten* pp.287–302.

Jacques de Clercq Information courtesy of drs Daan de Clercq, Amsterdam.

'A grower from Amsterdam ...' Krelage, 'Het manuscript over den tulpenwindhandel' pp.29–30.

Jan Breughel Wilfrid Blunt and William Stearn, *The Art of Botanical Illustration* p.128.

The Court of Holland and the resolution of the States Posthumus, 'De speculatie in tulpen' (1927) pp.56–60; Posthumus, 'The tulip mania in Holland in the years 1636 and 1637' p.146; Krelage, *Bloemenspeculatie in Nederland* p.93; Bulgatz pp.104–5.

In the event, the Court of Holland did hear at least one tulip case. This was a suit brought by the widow of Paulus van Beresteyn, who had been one of Haarlem's most eminent solicitors. Van Beresteyn came from a patrician family and was rich and influential enough to be counted among the regents of Haarlem even though he was a professed Catholic. He was a lieutenant of the civic guard and a governor of the Latin School, which prepared the children of the ruling class for university. He was an extremely wealthy man, with total capital well in excess of 12,000 guilders, and invested some of his money in Haarlem property. His interest in tulips, though, was probably that of a connoisseur rather than a florist. He lived in a large house on the Wijngaerderstraat and grew tulips in a garden on the Dijcklaan, a road which ran between two of the city's gates.

Van Beresteyn died, aged forty-eight, at the height of the mania in December 1636, two months before tulip prices crashed and eight weeks after selling six beds of tulips lying in his garden to a consortium of buyers comprising a local bookseller, Theunis Cas, and a second man named Jan Sael. The sale had been concluded on 29 September, before bulb prices began their final catastrophic rise, and the consortium paid the bargain price of 312 guilders – plus an atlas from Cas's shop – for the beds. Shortly afterwards, van Beresteyn sold the whole of his garden, excluding the bulbs, to a local bleacher named Nicolaes van der Berge. Van de Berge then approached Cas and Sael and agreed to buy the tulips for a total of 362 guilders. The agreement was that van der Berge would take on the consortium's debt to van Beresteyn's estate and pay them, in addition, a premium of 50 guilders. On 6 February, the day after prices in Haarlem crashed, Cas and Sael went to a local notary to confirm their willingness to proceed with this transaction, stating that tulips remained highly prized elsewhere in Holland, and in the summer van de Berge took possession of the bulbs when they were lifted. He failed, however, to pay for them when settlement fell due, and eventually the van Beresteyn family took action, issuing proceedings against not only the bleacher but also Cas and Sael.

Why this case, of all cases, found its way before the Court of Holland remains unclear. But it contains several striking features. It shows how difficult it was to determine who owned the bulbs traded during the mania, even when the chain of ownership was relatively short and straightforward; evidently, even those who had owned tulips only temporarily could easily be caught up in the melee of claim and counter-claim. It also demonstrates that long after the tavern trade collapsed there were some among the ranks of the richer traders and the connoisseurs who believed tulips were still a potentially good investment. Civiele processtukken II B 44, records of the Court of Holland, Algemeen Rijks Archief, The Hague; Index to *Heerenboek* p.12, Municipal Archives, Haarlem; Posthumus, 'De speculatie in tulpen' (1927) p.82; E. A. van Beresteyn and W. F. del Campo Hartman, *Genealogie van het Geslacht van Beresteyn* pp.133–6, 219–22.

Resolutions of the cities of Holland Posthumus, 'De speculatie in tulpen' (1927) p.60.

Munting Abraham Munting, *Naauwkeurige Beschryving der Aardgewassen* p.911.

Van Bosvelt Resolution of 5 November 1637, *Aanteekeningen van C. J. Gonnet*, Municipal Archives, Haarlem; Bulgatz p.105.

Many contracts nullified Posthumus, 'De speculatie in tulpen' (1927) p.69.

Cases in Alkmaar Ibid (1934) p.240.

De Block Ibid (1927) pp.48–9.

Abraham de Goyer Ibid pp.65–7.

Hans Baert Ibid p.76.

Admirael and de Hooge Ibid p.68.

Willem Schonaeus As well as being a poor judge of tulips, Koster must have been something of an optimist; even after the crash in prices, he agreed to continue with the transaction and he paid his deposit – 820 guilders, about 12 per cent of the purchase price – as late as 25 May. By the autumn, though, he had evidently changed his mind about the wisdom of the agreement and defaulted, forcing Schonaeus to take action. Posthumus, 'De speculatie in tulpen' (1927) pp.71, 79. Willem Schonaeus (1600–67) lived in one of Haarlem's best-known houses, De Hoofdwacht on the Grote Markt. G. H. Kurtz, 'De geschiedenis van ons vereeningsgebouw de Hoofdwacht' pp.37–8.

Cases in Haarlem Cf. Posthumus, 'De speculatie in tulpen' (1927) pp.71, 79. *De Clerq* Ibid pp.77, 79.

Haarlem's court of arbitration Ibid p.80; Krelage pp.96–7; Bulgatz p.105.

Friend-makers Brereton pp.8–9, 22; Posthumus, 'De speculatie in tulpen' (1927) p.80; *Aanteekeningen van C. J. Gonnet*, Municipal Archives, Haarlem; Posthumus, 'De speculatie in tulpen' (1934) pp.239–40.

Dubbelden Posthumus Ibid (1927) pp.84–5.

Van Goyen's insolvency It is not clear why Van Goyen did not take advantage of the opportunity to settle his debts at 3.5 per cent, which would have meant paying only 30 guilders. Probably the regents of The Hague did not follow their colleagues in Haarlem in setting up an arbitration panel to settle local cases.

14: AT THE COURT OF THE TULIP KING

Many of the books that were consulted for chapter 3 were also useful here, particularly those of Philip Mansel and Barnette Miller. Surprisingly, there seems to be no good biography of Ahmed III, but accounts of his tulip fêtes appear in numerous secondary sources, many of which have been drawn on; the most original and useful were Arthur Baker, 'The cult of the tulip in Turkey', and Michiel Roding and Hans Theunissen (eds), *The Tulip: A Symbol of Two Nations*. The historical background has been taken both from general histories such as Alan Palmer, *The Decline and Fall of the Ottoman Empire* and more specialist studies, including Lavender Cassels, *The Struggle for the Ottoman Empire, 1717–1740.*

Mehmed IV and the tulip Palmer pp.10, 14–15, 37; Turhan Baytop, 'The tulip in Istanbul during the Ottoman period', in Roding and Theunissen pp.50–6; Barnette Miller, *Beyond the Sublime Porte* p.124.

Ibrahim the Mad During his eight-year reign, he was also noted for deflowering a virgin every Friday. Palmer p.19; Norman Penzer, *The Harem* pp.188–91.

The Time of Tulips Fatma Müge Göçek, *East Encounters West: France and the Ottoman Empire in the Eighteenth Century* p.10.

Nedim the poet Palmer p.36; Andrew Wheatcroft, *The Ottomans* pp.77, 79; Mansel p.181.

'execution might after all still be their lot ...' When court officials entered the cage to call Suleyman II (1687–91) to the throne in succession to Mehmed IV, the new sultan is said to have cried out in terrified exasperation: 'If my death has been commanded, say so. Since my childhood, I have suffered forty years of imprisonment. It is better to die at once than to die a little every day. What terror we endure for a single breath.' Halil Inalcik, *The Ottoman Empire* p.60.

Sultan Ahmed's flower festivals Noel Barber, *The Lords of the Golden Horn: From Suleiman the Magnificent to Kamal Ataturk* pp.109–10; Mansel pp.76–8, 180–1; Palmer pp.37–8; Miller pp.124–6; Penzer pp.258–60.

General passion for tulips in Ahmed's reign Yildiz Demiriz, in Roding and Theunissen pp.57–8; Baytop pp.55; Baker p.235.

Eighteenth-century criteria for ideal tulips Baytop p.53; Demiriz pp.57–8; W. S. Murray, 'The introduction of the tulip, and the tulipomania' p.20.

Ottoman officials' flowers and bribes of tulips Mansel p.182; Stanford Shaw, *History of the Ottoman Empire and Modern Turkey* p.234.

Fazil Pasha Mansel p.147.

Damat Ibrahim Palmer pp.33–5, 38.

The Sa'adabad Ibid p.34; Shaw p.234; Mansel pp.180–1; Göçek pp.51, 79; Alexander Pallis, *In the Days of the Janissaries* p.199.

The fall of Damat Ibrahim and Ahmed III Palmer pp.38–9.

Mahmud I and the decline of the tulip in Turkey Barber p.110; Wheatcroft pp.80–1.

15: LATE FLOWERING

The later history of the bulb trade is reliably covered in modern histories. The hyacinth trade is described in detail by E. H. Krelage in *Bloemenspeculatie in Nederland: De Tulpomanie van 1636–37 en de Hyacintenhandel 1720–36*, and the later history of the tulip by both Krelage, in *Drie Eeuwen Bloembollenexport*, and Daniel Hall, in *The Book of the Tulip*.

Continuing trade in tulips Krelage, *Bloemenspeculatie in Nederland* pp.97–110; Krelage, *Drie Eeuwen Bloembollenexport* pp.15–18; Sam Segal, *Tulips Portrayed* p.17;

Peter Mundy, *The Travels of Peter Mundy*, vol.4 p.75; Peter Garber, 'Tulipmania' pp.550–3.

Aert Huybertsz. Nicolaas Posthumus, 'De speculatie in tulpen in de jaren 1636 en 1637' (1927) pp.82–3.

Haarlem as the centre of the later bulb trade Krelage, *Bloemenspeculatie in Nederland* pp.102–4; *Drie Eeuwen Bloembollenexport* pp.9–11.

Desiderata of van Oosting and van Kampen Cited by Segal p. 11 and Hall pp.48–9.

The hyacinth trade Krelage, *Bloemenspeculatie in Nederland* pp.142–96; Krelage *Drie Eeuwen Bloembollenexport* pp.13, 645–55; Garber pp.553–4; Joseph Bulgatz, *Ponzi Schemes, Invaders from Mars* pp.109–14.

The history of the tulip to the present day Krelage, *Drie Eeuwen Bloembollenexport* pp.15–18.

Craze for dahlias Bulgatz pp.108–9. During this episode, there was even talk of the propagation of blue dahlias – as much a botanical impossibility as the black tulip.

Craze for gladioli Posthumus, 'The tulip mania in Holland in the years 1636 and 1637' p.148.

Chinese spider-lily mania Burton Malkiel, *A Random Walk Down Wall Street* pp.82–3.

Coca-Cola auction. Mark Pendergrast, *For God, Country and Coca-Cola* p.211.

Bibliography

1. UNPUBLISHED MATERIAL

Municipal Archives, Haarlem
Notarial registers, vols. 120–50
Burial registers, vols. 70–6
Index to *Heerenboek*
Manuscript entitled *Aantekeningen van C. J. Gonnet Betreffende de Dovestalmanege in de Grote Houstraat, de Schouwburg op het Houtplein, het Stadhuis in de Frase Tijd, Haarlemse Plateelbakkers en Plateelbakkerijen en de Tulpomanie van 1637–1912*

Stadsbibliotheek, Haarlem
Chrispijn van de Passe, *Een Cort Verhael van den Tulipanen ende haere Oefeninghe ...* (contemporary pamphlet, np, nd)

Municipal Archives, Amsterdam
Burial registers

Algemeen Rijks Archief, The Hague
Records of the Court of Holland

Posthumus Collection, Netherlands Economic History Archive
Copies of unpublished acts relating to the tulip mania from the Notarial Archives of Alkmaar and Leiden

2. PUBLISHED MATERIAL

Aitzema, Lieuwe van *Saken van Staet en Oorlogh* (vol.II, 1633–44) (The Hague: Johan Veely, Johan Tongerloo & Jasper Doll, 1669)

Baker, Arthur 'The cult of the tulip in Turkey', *Journal of the Royal Horticultural Society*, September 1931

Barber, Noel *The Lords of the Golden Horn: From Suleiman the Magnificent to Kamal Ataturk* (London: Macmillan, 1973)

Beijer, T., et al *Nicolaes Tulp. Leven en Werk van een Amsterdamse Geneesheer en Magistraat* (Amsterdam: Six Art Promotion, 1991)

Beresteyn, E. A. van and W. F. del Campo Hartman, *Genealogie van het Geslacht van Beresteyn* (The Hague: np, 1941 and 1954)

Blunt, Wilfrid *Tulipomania* (London: Penguin, 1950)

Blunt, Wilfrid, and William Stearn *The Art of Botanical Illustration* (Woodbridge: The Antique Collectors Club, 1994)

Boer, H. W. J. de, H. Bruch et al *Adriaan Pauw (1585–1653): Staatsman en Ambachtsheer* (Heemstede: Vereniging Oud-Heemstede-Bennebroek, 1985)

Boxhornius, Marcus Zuerius *Toneel, ofte Beschrijvinghe des Landts, ende Steden van Hollandt ende West-Vrieslandt* (Amsterdam: Hendrik Hondius, 1632)

Brereton, William *Travels in Holland, the United Provinces etc ... 1634–1635* (London: Chetham Society, 1844)

Bulgatz, Joseph *Ponzi Schemes, Invaders from Mars and More Extraordinary Popular Delusions and the Madness of Crowds* (New York: Harmony, 1992)

Carswell, John *The South Sea Bubble* (Stroud: Alan Sutton, 1993)

Cassels, Lavender *The Struggle for the Ottoman Empire, 1717–1740* (London: John Murray, 1966)

Cos, P. *Verzameling van een meenigte tulipaanen, naar het leven geteekend met hunne naamen, en swaarte der bollen, zoo als die publicq verkogt zijn, te Haarlem in den jaare A.1637, door P. Cos, bloemist te Haarlem* (Haarlem: np, 1637)

Cotterell, Geoffrey *Amsterdam: The Life of a City* (Farnborough: D. C. Heath, 1973)

Damme, A. van *Aanteekeningen Betreffende de Geschiedenis der Bloembollen: Haarlem 1899–1903* (Leiden: Boerhaave, 1976)

Deursen, A. T. van *Plain Lives in a Golden Age: Popular Culture, Religion and Society in Seventeenth Century Holland* (Cambridge: Cambridge University Press, 1991)

Dijk, W. van *A Treatise on Tulips by Carolus Clusius of Arras* (Haarlem: Enschedé, 1951)

Dumas, Alexandre *The Black Tulip* (Oxford: Oxford University Press, 1993)

Eeghen, I. H. van 'Een oude band met gedichten: Gerret Jansz. Kooch' in *Maandblad Amstelodamum* 53 (1966)

Ehrenberg, R. *Grosse Vermögen* (Jena: Gustav Fischer, 1925)

Evelyn, John *The Diary of John Evelyn*, II, Kalendarium 1620–1649 (Oxford: Clarendon Press, 1955)

Fischer, Hans *Conrad Gesner 1516–1565. Leben und Werk* (Zürich: Leemann, 1966)

Garber, Peter M. 'Tulipmania', *Journal of Political Economy* 97 (I), June 1989, pp.535–60

Gelder de Neufville, D. M. van 'De oudste generaties van het geslacht de Neufville' in *De Nederlandsche Leeuw* (1925)

Geyl, Pieter *The Revolt of the Netherlands 1555–1609* (London: Cassell, 1988)

Göçek, Fatma Müge *East Encounters West: France and the Ottoman Empire in the Eighteenth Century* (New York: Oxford University Press, 1987)

Goodwin, Jason *Lords of the Horizon: A History of the Ottoman Empire* (London, Chatto & Windus, 1998)

Griffey, E. 'What's in a name? Forging an identity: portraits of Nicholaes Tulp (1593–1674)' in *Dutch Crossing* 21 (1997) pp.3–43

Groenveld, S., E. K. Grootes, J. J. Temminck et al, *Deugd Boven Geweld. Een Geschiedenis van Haarlem 1245–1995* (Hilversum: Verloren, 1995)

Hall, A. Daniel *The Book of the Tulip* (London: Martin Hopkinson, 1929)

Harline, E. Craig *Pamphlets, Printing and Political Culture in the Early Dutch Republic* (Dordrecht: Martinus Nijhoff, 1987)

't Hart, Marjolein, Joost Jonker and Jan Luiten van Zanden (eds) *A Financial History of the Netherlands* (Cambridge: Cambridge University Press, 1997)

Hensen, A. 'De vereering van St Nicholaas te Amsterdam in den Roomschen tijd' in *Bijdragen voor de Geschiedenis van het Bisdom Haarlem*, 43 (Haarlem, 1925) pp.187–91

Herbert, Zbigniew *Still Life with a Bridle* (London: Jonathan Cape, 1993)

Hondius, Petrus *Dapes Inemptae, of de Moufe-schans, dat is, De soeticheydt des buytenlevens, vergheselschapt met de boecken* (Leiden: Daniel Roels, 1621)

Hunger, F. W. T. *Charles d'Escluse (Carolus Clusius), Nederlandsche Kruidkundige 1526–1609*, 2 vols (The Hague: Martinus Nijhoff, 1927 and 1943)

Inalcik, Halil *The Ottoman Empire: The Classical Age 1300–1600* (London: Phoenix, 1994)

Israel, Jonathan *The Dutch Republic: Its Rise, Greatness and Fall, 1477–1806* (Oxford: Oxford University Press, 1998)

Jacob, Joseph *Tulips* (London: J. C. & E. C. Jaek, 1912)

Kindleberger, Charles *Manias, Panics and Crashes: A History of Financial Crises* (New York: John Wiley, 1996)

Krelage, E. H. *Bloemenspeculatie in Nederland: De Tulpomanie van 1636–37 en de Hyacintenhandel 1720–36* (Amsterdam: Kampen, 1942)

——*De Pamfletten van den Tulpenwindhandel 1636–1637* (The Hague: Martinus Nijhoff, 1942)

——'Het manuscript over den tulpenwindhandel uit de verzameling Meulman', *Economisch-Historisch Jaarboek* XXII (1943)

——*Drie Eeuwen Bloembollenexport* (The Hague: Rijksuitgeverij, 1946)

Kurtz, G. H. 'De geschiedenis van ons vereenigingsgebouw de Hoofdwacht' in *Jaarboek Haarlem* (1942) pp.32–52.

——'Twee oude patriciërshuizen in de Kruisstraat' in *Jaarboek Haarlem* (1961) pp.112–42.

Leonhardt, G. *Het Huis Bartolotti en zijn Bewoners* (Amsterdam: Meulenhoff, 1979)

Lesger, C. and L. Noordegraaf (eds) *Entrepreneurs and Entrepreneurship in Modern Times: Merchants and Industrialists Within the Orbit of the Dutch Staple Market* (The Hague, 1995)

Mackay, Charles *Memoirs of Extraordinary Popular Delusions and the Madness of Crowds* (Ware: Wordsworth Editions, 1995)

Malcolm, Noel *Kosovo: A Short History* (London: Macmillan, 1998)

Malkiel, Burton *A Random Walk Down Wall Street* (New York: W. W. Norton, 1996)

Mansel, Philip *Constantinople: City of the World's Desire, 1453–1924* (London: John Murray, 1995)

Martels, Z. R. M. W. von *Augerius Gislenius Busbequius: Leven en Werk van de Keizerlijke Gezant aan het hof van Süleyman de Grote* (University of Groningen, 1989)

Mather, John *Economic Production of Tulips and Daffodils* (London: Collingridge, 1961)

Miller, Barnette *Beyond the Sublime Porte: The Grand Seraglio of Stambul* (New Haven: Yale University Press, 1931)

Moryson, Fynes *An Itinerary Containing His Ten Yeeres Travell Through the Twelve Dominions of Germany, Bohmerland, Sweitzerland, Netherland, Denmarke, Poland, Italy, Turkey, France, England, Scotland and Ireland* (4 vols, Glasgow: James MacLehose & Sons, 1907)

Mundy, Peter *The Travels of Peter Mundy* (4 vols, London: Hakluyt Society, 1907–24)

Munting, Abraham *Waare Oeffening der Planten* (Amsterdam: Hendrik Rintjes, 1671)

——*Naauwkeurige Beschryving der Aardgewassen* (Leiden: Pierre Van der Aa, 1696)

Murray, W. S. 'The introduction of the tulip, and the tulipomania', *Journal of the Royal Horticultural Society*, March 1909 pp.18–30.

Nieuw Nederlandsch Biographisch Woordenboek vols IV, V, VI, VII (Leiden: A. W. Sijthoff, 1918, 1921, 1924, 1930)

Norwich, John Julius *Byzantium: The Decline and Fall* (London: Viking, 1995)

Pallis, Alexander *In the Days of the Janissaries: Old Turkish Life as Depicted in the 'Travel-Book' of Evliyá Chelebí* (London: Hutchinson, 1951)

Palmer, Alan *The Decline and Fall of the Ottoman Empire* (London: John Murray, 1992)

Parker, Geoffrey *Europe in Crisis 1598–1648* (London: Fontana, 1979)

Pavord, Anna *The Tulip* (London: Bloomsbury, 1998)

Pendergrast, Mark *For God, Country and Coca-Cola* (New York: Touchstone, 1997)

Penzer, Norman *The Harem: An Account of the Institution as it Existed in the Palace of the Turkish Sultans, with a History of the Grand Seraglio from its Foundation to Modern Times* (London: Spring Books, 1966)

Polnitz, G. Freiherr von *Die Fugger* (Tübingen: J. C. B. Mohr, 1981)

Posthumus, Nicolaas 'De speculatie in tulpen in de jaren 1636 en 1637' parts 1–3, *Economisch-historisch jaarboek* 12 (1926) pp.3–99; 13 (1927) pp.1–85; 18 (1934) pp.229–40.

——*Inquiry into the History of Prices in Holland* (2 vols, Leiden: E. J. Brill, 1946)

——'The tulip mania in Holland in the years 1636 and 1637', in W. C. Scoville and J. C. LaForce (eds), *The Economic Development of Western Europe* vol.II (Lexington, Mass., 1969)

Price, J. L. *Culture and Society in the Dutch Republic During the 17th Century* (London: B. T. Batsford, 1974)

Raaij, H. L. Janssen van *Kroniek der Stad Haarlem van de Vermoedelijke Stichting der Stad tot het Einde van het Jaar 1890* (Haarlem: Loosjes, c.1894)

Regteren Altena, L. Q. van *Jacques de Gheyn: Three Generations* vol.1 (The Hague: Martinus Nijhoff, 1983)

Roding, Michiel, and Hans Theunissen (eds), *The Tulip: A Symbol of Two Nations* (Utrecht and Istanbul: Turco-Dutch Friendship Association, 1993)

Rohde, Eleanour *Crispian Passeus's 'Hortus Floridus'* (London, 1928–9)

Savage, Spencer 'The "Hortus Floridus" of Crispijn van de Pas', *Transactions of the Bibliographic Society*, Series II, 4 (1923) pp.181–206

——*Hortus Floridus: The Four Books of Spring, Summer, Autumn and Winter Flowers, Engraved by Crispin van de Pas* (London: Minerva, c.1974)

Schama, Simon *The Embarrassment of Riches: An Interpretation of Dutch Culture in the Golden Age* (London: Fontana, 1991)

Schloredt, Valerie *A Treasury of Tulips* (London: Michael O'Mara Books, 1994)

Schrevelius, Theodorus *Harlemias of Eerste Stichting der Stad Haarlem* (Haarlem: Johannes Marshoorn, 1754)

Segal, Sam *Tulips by Anthony Claesz: 56 Seventeenth-Century Watercolour Drawings by Anthony Claesz (ca.1607/ 8–1648)* (Maastricht: Noortman, 1987)

——*Tulips Portrayed: The Tulip Trade in Holland in the Seventeenth Century* (Lisse: Museum voor de Bloembollenstreek, 1992)

Segal, Sam and Michiel Roding, *De Tulp en de Kunst. Verhaal van een Symbool* (Zwolle: Waanders, 1994)

Shaw, Stanford *History of the Ottoman Empire and Modern Turkey* (2 vols, Cambridge: Cambridge University Press, 1976)

Slikke, C. M. van der *Tulpenteelt op Kleigrond* (Berlikum, 1929)

Slogteren, E. van 'Broken tulips', in *The Daffodil and Tulip Yearbook* (London: Royal Horticultural Society, 1960)

Solms-Laubach, Hermann, Grafen zu *Weizen und Tulpe und deren Geschichte* (Leipzig: Arthur Felix, 1899)

Stoye, John *English Travellers Abroad, 1604–1667* (New York: Octagon Books, 1968)

Taylor, Paul *Dutch Flower Painting 1600–1720* (London: Hale, 1995)

Temminck, J. J. 'Naar haer spraecke gebooren van Amsterdam. Enkele gegevens over de relatie tussed Haarlem en Amsterdam in vroeger eeuwen', in *Jaarboek Haarlem* (1981) pp.43–67.

Temminck, J. J. et al *Haarlemmerhout 400 Jaar. 'Mooier is de Wereld Nergens.'* (Haarlem: Schuyt & Co., 1984)

Temple, William *Observations Upon the United Provinces of the Netherlands* (Cambridge: Cambridge University Press, 1932)

Theunisz, Johan *Carolus Clusius: Het Merkwaardige Leven van een Pionier der Wetenschap* (Amsterdam: P. N. Van Kampen & Zoon, 1939)

Vogelaar, C. *Jan van Goyen* (Zwolle: Waanders, 1996)

Vries, Jan de *The Dutch Rural Economy in the Golden Age, 1500–1700* (New Haven: Yale University Press, 1974)

Vries, Jan de, and Ad van der Woude, *The First Modern Economy: Success, Failure and Perseverance of the Dutch Economy 1500–1815* (Cambridge: Cambridge University Press, 1997)

Wassenaer, Nicolaes Jansz. van *Historisch Verhael aller Gedencwaerdiger Gheschiedenissen,* V–IX (Amsterdam: Iudocus Hondius and Jan Jansen, 1624–5)

Watt, Tessa *Cheap Print and Popular Piety, 1550–1640* (Cambridge: Cambridge University Press, 1991)

Weider, E. C. 'De pamflettenverzameling van den Amsterdammer Abraham de Goyer van 1616', in *Het Boek* 6 (The Hague: Martinus Nijhoff, 1917)

Werner, J. W. K. *Haarlemmermeer. 17e en 18e Eeuwse Voorstellen tot Droogmaking* (Amsterdam: np, 1985)

Wheatcroft, Andrew *The Ottomans: Dissolving Images* (London: Penguin, 1993)

Whiteway, R. S. *The Rise of Portuguese Power in India 1497–1550* (London: Archibald Constable, 1899)

Wijnands, O. 'Tulpen naar Amsterdam: plantenverkeer tussen Nederland en Turkije', in H. Theunissen, A. Abelman and W. Meulenkamp, *Topkapi en Turkomanie: Turks-Nederlandse Ontmoetingen Sinds 1600* (Amsterdam: De Bataafsche Leeuw, 1989)

Zumthor, Paul *Daily Life in Rembrandt's Holland* (London: Weidenfeld & Nicolson, 1962)

Acknowledgements

Of all the many debts of gratitude incurred during the writing of this book, the greatest is certainly that I owe to my indefatigable research assistant, drs Henk Looijesteijn of Amsterdam. Being both a specialist in early modern history and himself the son of a long line of bulb growers, drs Looijesteijn could hardly have been better qualified to conduct original research on my behalf in the archives of Haarlem, Amsterdam and The Hague and to serve as my guide to the copious Dutch literature on the subject. *Tulipomania* could not have been written without him.

My introduction to drs Looijesteijn came through the courtesy of Henk van Nierop of the University of Amsterdam, himself among the most distinguished of the historians of the period. Others who assisted me in the Netherlands included Jaap Looijesteijn, bulb grower of Breezand, and drs Daan de Clercq of Amsterdam, who shared information concerning his ancestor Jacques de Clercq.

I am grateful for the help and guidance of my agent, Patrick Walsh, and my editor, Sara Holloway, who had faith in the idea and salvaged what they could from my confused initial thoughts upon the subject. Tina Walsh translated some particularly obscure passages of Old Dutch for me. The person who worked longest and hardest to see this book to completion, however, was Penny, who has my thanks and all my love. This book is hers as well as mine.

Index